Royal Botanic Gardens **Kew**

THE GARDENER'S COMPANION TO

PESTS & DISEASES

A GUIDE TO DIAGNOSIS,
TREATMENT & PREVENTION

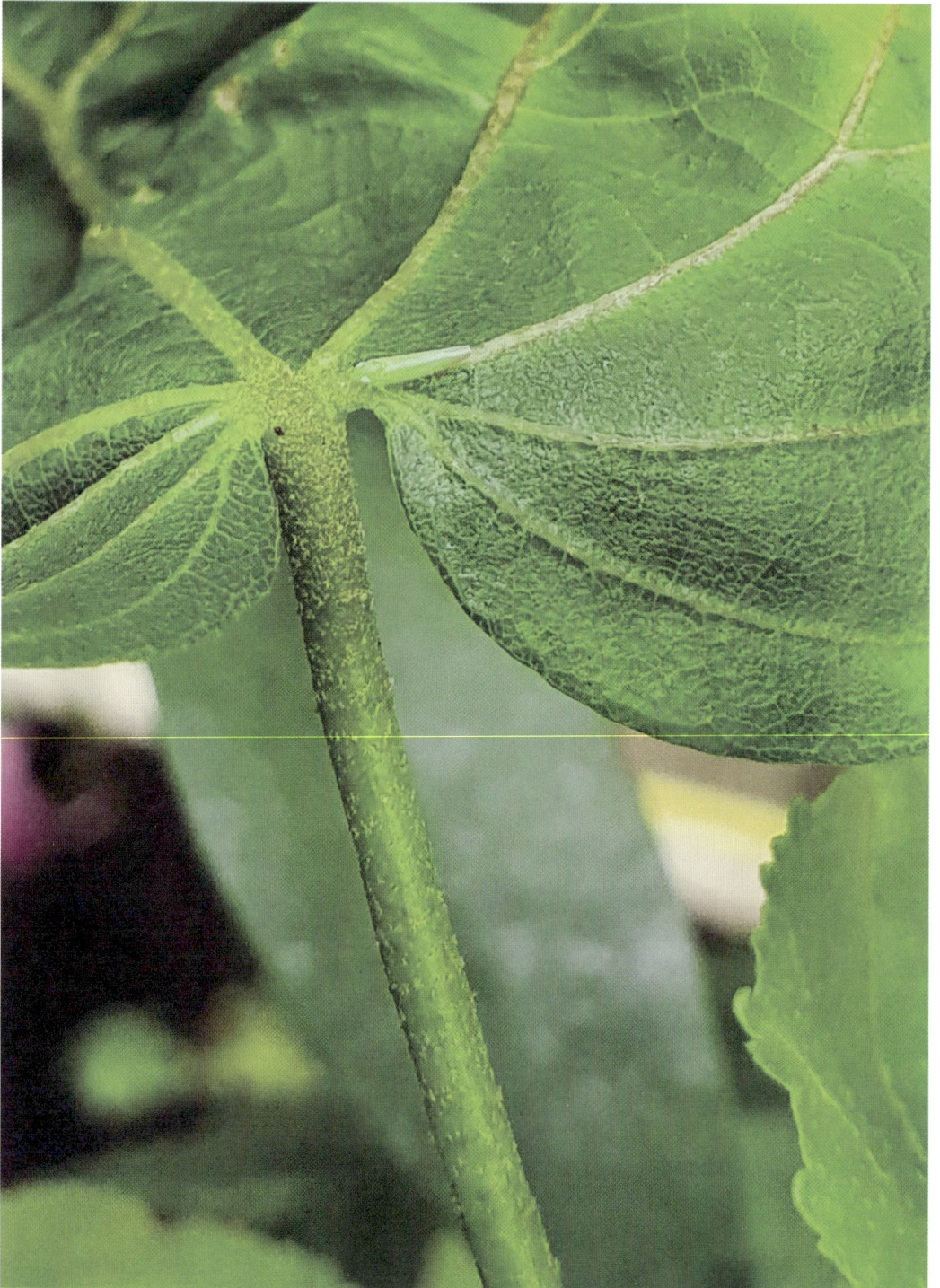

Royal Botanic Gardens **Kew**

THE GARDENER'S COMPANION TO

PESTS & DISEASES

A GUIDE TO DIAGNOSIS,
TREATMENT & PREVENTION

Hélèna Dove, Polly Stevens,
Kevin Martin and Paul Rees

F FRANCES
LINCOLN

Contents

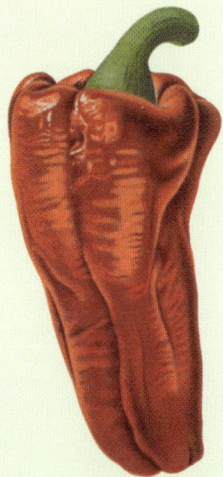

INTRODUCTION

Gardens are a haven in which we can sit, cultivate and collect our thoughts. But among the tranquillity of this lush green haven, an ongoing war wages between the gardener and the pests and diseases that also want to set up camp in our gardens. In days gone by, these issues were dealt with primarily by chemicals, but this solution has proved to cause imbalances in the ecosystems, doing more harm than good. New ways of managing pests and diseases have subsequently developed; these involve working in harmony with nature.

At the Royal Botanic Gardens, Kew, we use sustainable methods where possible to manage our plant collections, and in this book we aim to pass this knowledge on to the home gardener. Understanding what brings a pest or disease to your garden is key – can you avoid them visiting in the first place? And if they do arrive, can you tolerate the damage or is action essential? Can you mitigate the issues by creating a diverse, wildlife-friendly garden? The horticultural industry has long practised Integrated Pest Management (IPM), where the aim is to use a range of techniques to monitor, deter and manage pests and diseases, rather than rely on one tool alone, and many of these techniques are described in this book.

Most pests and diseases are already present in the wider environment, but by creating a garden packed full of plants we are providing spaces that will naturally attract insects looking for a tasty lunch as well as diseases that need a host.

WHAT ARE GARDEN PESTS?

The term 'pest' is bandied around a fair amount among gardeners, but what are we actually talking about when we mutter this under our breath at the massive holes in our *Dahlia* leaves? In the context of horticulture, a pest is an animal that causes damage to plants and comes in a whole range of guises, from birds to deer and from aphids to slugs. Pests differ wildly from garden to garden in whether they appear outdoors or under cover. Only a small number of animals are pests.

Below, left: Beneficial insects including some wasps will eat pests like aphids.
Below, right: Ants will farm and protect mealybugs, so an easy way to manage mealybugs can be to deter the ants.
Right: Having a mixture of flower forms, such as these *Tagetes* and the ornamental carrot (*Daucus*) in the centre, attracts a wide variety of beneficial insects.

Outdoors, it is quite common for the pest to be an animal that already inhabits the area, such as vine weevil grubs (see page 73) in a hedgerow or cabbage white butterflies (see page 36) in grassland. In the wild, these pests generally feast on species plants, which tend to be more resistant to damage but also are not as tasty as our cultivated garden plants. All it takes is for a nearby gardener to fill their plot with some delicious new plants – and not with just the odd one or two but whole swathes of them – the pests will then hurry over to try this fine dining experience. In the case of these examples, vine weevils would much rather eat the tender roots of potted plants, where they will be much less disturbed than in their hedgerow habitats.

Although pests occur naturally, some do hitch a ride into the growing space via plants, seeds or even natural materials such as wood, especially if these items haven't undergone proper inspections (see page 16) or treatments, for example International Standards for Phytosanitary Measures No. 15 (ISPM 15). Many new pests have become resident in our gardens, especially as climate change creates conditions where more pests can thrive.

Below: Rotting wood left around the garden provides habitats that support many beneficial animals and insects, such as the millipede pictured below.

Above: Vine weevils naturally occur in hedgerows but become a pest when they take up residence in plant pots and other containers.

One of the key characteristics of some pests is their ability to reproduce fast, and nowhere can this be seen more vividly than when growing plants under cover, either indoors or in a greenhouse. The warm, relatively humid and consistent environment that we like to create for ourselves to live in enables these pests to reproduce and cause damage at an alarming rate. Indoor pests also fear very little, as there are few in the way of predators, so the gardener needs to be extremely vigilant and act fast. As part of IPM, understanding the life cycles of pests helps to target them when most vulnerable.

WHAT ARE GARDEN DISEASES?

While pests are a pain, and a plague of small insects isn't pleasant, plant diseases tend to come from the inside of the plant and develop outwards – even the most vigilant gardener can be taken by surprise. Diseases and pests often work together, with pests causing physical damage to a plant, thereby enabling a disease to enter.

The term garden disease covers a multitude of issues, from infectious organisms to physiological factors. Predominantly, when talking about garden diseases, we are describing those caused by pathogens such as bacteria, viruses, fungi and protists. Garden diseases can show themselves in many ways, and one disease can have several symptoms such as curled leaves, necrotic patches, oozing wounds and various other grim physical ailments. Physiological diseases are triggered by numerous environmental factors including weather and nutrient deficiencies.

For a disease to occur, a 'disease triangle' of three elements are in play: the right environment, the pathogen and the host plant. Removing any of these elements would completely stop the disease, but all are hard to completely irradicate, so improving all three helps the situation. Adjusting the environment to a balance in which the host can survive but the pathogen struggles will weaken the pathogen: for example, fungal infections require water to reproduce and, although plants require water, you can water at the base of the plant, which should keep leaves dry when fungal diseases such as black spot on roses (see page 82) are rife.

Below: Pear rust flourishes in humid summers, but thankfully causes little real damage. Discarding fallen leaves is the best way to control its spread.

Above, left: Powdery mildew affects the leaves of many plants and thrives in humid conditions. To help reduce outbreaks, encourage good air flow by thinning plants and reducing foliage.

Above, right: Rusts are caused by fungi and are common both under glass and outdoors. They rarely kill a plant, but destroying infected material slows their spread.

Eliminating the pathogen is hard, but removing and destroying infected material will slow it in its tracks. Never compost diseased material as most domestic compost heaps do not reach a high enough temperature to kill off pathogens and may contain even more material suitable for the sneaky diseases. Instead, send diseased material to the local composts, where the material is cooked to a higher temperature, or burn it in a home incinerator if allowed in your growing space. The last element from the disease triangle is the host plant. Logically, we are trying to save the plant and do not want to remove it from play, but it is worth assessing any plant before introducing it to the garden or home. If you know there is a disease in the area, should a conscientious gardener be planting a potential host plant that will help the disease spread? For example, if you know that your neighbour struggles with fireblight (see page 78) on their firethorn (*Pyracantha*) should you plant a firethorn just several metres away, when it will inevitably struggle?

HOW TO IDENTIFY A PROBLEM

Keeping a close eye on your plants is the best way to know if they are happy and thriving or under attack from a pest or disease. If caught early, most pests and diseases can be managed and our plants are free to live on. But how do you know what the problem is, as so many pests and diseases present with the same symptoms?

Always visualize what a healthy plant specimen looks like so that you can compare. Are its leaves normally lime-green, or should they be a dark green? Does the bark on this tree crack off naturally, or is the peeling abnormal? Should the flower stem be 10cm/4in high, or does it normally grow to 1m/3ft? If you don't know offhand, ask at a garden centre or nursery, visit gardens or consult the internet, where there are always useful guides and descriptions. Keeping a good log of photos helps to remind you of how a plant should perform and, when it doesn't match the descriptor, should give you a good start to identifying the issue.

Above: A great way to monitor plant health is to stroll through the garden removing small infestations of pests by hand or infected plant material. This can often stop a disaster.

Chatting to other gardeners and reading gardening books are great ways to familiarize yourself with the possible pests and diseases that your plants may encounter, and having this information stored away in the back of your mind helps with a swift diagnosis. Sadly, there are also new pests and diseases making their way into growing spaces, and to keep up to date with these it is worth checking online sources such as the UK Plant Health Information Portal run by the Department for Environment, Food & Rural Affairs.

If pests are the suspected problem, a hand lens is a useful tool to get a closer view of whatever is destroying the plant. Remember that insects undergo many instars (stages) and may look incredibly different along the way – a caterpillar transforming into a butterfly being one of the most familiar. How the pest is affecting the plant can also help identify the scoundrel and put an end to its damage. Is the pest mainly on young fresh growth, sucking away at sap? Is it eating away at the edges of the foliage, leaving behind notched patterns?

Diseases can be trickier to pinpoint as often the symptoms look similar when viewed with the naked eye. In extreme cases, send material in a tightly sealed bag away to laboratories that will analyze the issue and advise on how to proceed.

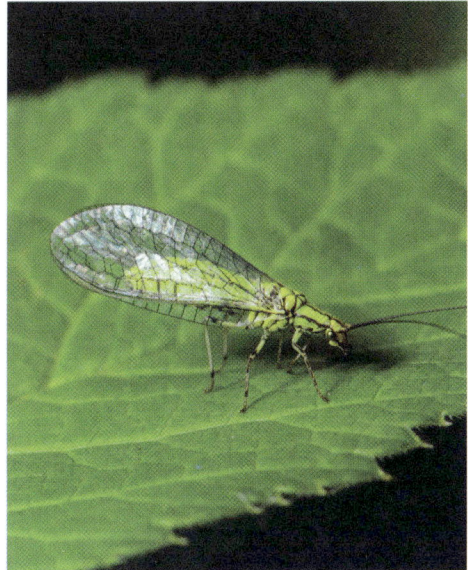

Below, left: Asparagus beetles lay their eggs along the stems of the asparagus plant, and the resulting offspring will strip the plant of its foliage. Hand removal of the adults and eggs is one of the most efficient ways of controlling this pest. *Below, right*: Lacewing larvae can eat a large variety of pests such as aphids, mealybugs and spider mites. Attract the adult lacewing to the garden by choosing plants, such as sunflowers (*Helianthus annuus*) and *Cosmos*, with single open flowers.

HOW TO PREVENT PESTS AND DISEASES IN YOUR GARDEN

Ideally, you can stop pests and diseases before they attack your beautiful and bountiful growing spaces. There are a multitude of tricks you can employ even before planting, which can help stop disaster in its tracks.

Firstly, and possibly most importantly, always look at the label on your plant or seeds. They should all have a plant passport if coming from a reputable source. This means the nursery is working to the highest standards, and if they did experience a pest or disease outbreak the staff could trace their plants and inform buyers, and vice versa. Seed potatoes, for example, have to be certified free of at least twenty pests and diseases.

Secondly, a visual inspection before buying a plant also helps. Does the plant look healthy to you? Is there anything wiggling away in the soil? Are the leaves an even colour? Quite often your instincts are right when it comes to how a plant presents, and you can spot an unhealthy specimen fairly easily.

Below: Look out for plant passport labels. These contain information such as the country of origin, nursery registration number and the traceability number.

Above: Protecting plants before pests can even get the chance to attack is one of the most environmentally friendly techniques you can adopt.

Thirdly, once a plant makes it to your home, you will want to stop pests and diseases from reaching it. With diseases, it is a little tricky, as they can travel on many vectors including the wind, water, insects and even on your boots. One easy way to manage disease in the garden is to keep tools clean, wiping them down after use and cleaning dirt off (see How to clean tools and other equipment, page 104).

Finally, stopping pest attacks can be tricky, but often a barrier is enough to slow their progress. Particularly in the vegetable patch, different types of netting can stop caterpillars, birds and mammals (see How to protect plants from birds, rabbits and other mammals, page 74). Gentle patrols around the garden, with an eye to collecting and managing some pests, will stop infestations. Also, remove hiding spaces to reduce the pests taking up residence, as they want somewhere to hide from predators; taking away pots, clearing leaves and tidying safe little nooks will help.

WILDLIFE-FRIENDLY SOLUTIONS

Entirely preventing pests and diseases often creates a sterile growing space, which is where wildlife-friendly gardening and a balanced approach come in. A small patch of lawn dug up by a mole, while annoying, isn't the end of the world, and provides some very nice seed compost; while a little pear rust is frustrating but doesn't kill off the tree, instead just lowers its productivity, so just remove the fallen leaves rather than attack the tree itself.

Wildlife-friendly gardening includes attracting or even introducing beneficial insects into your space, which will hopefully feast on the pests that attack the plants you want to grow. There are simple ways of doing this, which will make space attractive, too. Adding a water feature will help support many beneficial species, such as frogs and toads, which love to snack on slugs (see page 149). Water will bring birds such as sparrows and wrens to the garden, too, as they will use it to drink and wash in, while also finding a snack of tasty aphids (see page 134).

Below: Although moles can destroy the aesthetics of a formal lawn, the soil they dig up makes a fantastic medium in which to grow seeds, and they aerate the ground really well.

Above: Small birds are fantastic in the garden, eating pests such as small slugs and aphids. To attract them, introduce a bird bath, which will give the birds somewhere to wash and drink.

Planting to attract beneficial insects is not only practical but also good fun and beautiful. To get a range of insects, you need a variety of flower colours and shapes: for example, lacewing larvae are some of the best aphid eaters out there, and the parents love an umbel where the nectar is easily accessible, so plant fennel (*Foeniculum*) or ornamental carrots (*Daucus*) for these. Ladybird adults and larvae also enjoy an aphid, alongside pollen from plants such as nettles (*Urtica*), so finding a corner for these prickly greens is a must, as they support a lot of other beneficial insects as well. Interplanting will help protect crops; a row of *Dahlia* is extremely easy to find, whereas when these plants are hidden among roses (*Rosa*) and sage (*Salvia*) they are a little harder for the pests to spot (see How to use companion planting, page 52).

Predatory beetles will eat slugs, caterpillars and other soft-bodied pests, but to give them a home they need a corner of decaying wood, which, again, will be welcomed by many other beneficial species, thereby increasing the biodiversity of the garden.

One note on wildlife gardening is that a little damage to plants needs to be tolerated. For beneficial insects to make a home in your garden, a constant supply of their food should be available, so if all the aphids disappear then the lacewings will fly elsewhere to find their meal and not be present when the next outbreak occurs. Thus, when a few aphids are on your plants, leave them to be eaten for lunch by someone else in the food chain – as long as the aphids aren't causing big problems.

Beneficial organisms can also be introduced if not already present: for example, nematodes, which manage slug populations, can be watered on to your plot and into pots. Such beneficial insects can be great for indoor growing, where there are few natural predators, but, again, balance is required. One of the best biological controls for whitefly (see page 166) is a tiny parasitic wasp *Encarsia formosa*. The female lays her eggs in the whitefly, and the resulting larvae consume the whitefly from the inside. This wasp reproduces asexually, producing only female offspring until there are a high number of parasitized whitefly, then they produce males. Thus, to keep a steady stream of female wasps, they should be introduced slowly, over time,

Below: Frogs are a great addition to any garden, as they consume small slugs and snails. Give them a place to live, such as a pond, and somewhere to shelter, and they will pay you dividends.

Above, left: An parasitic wasp called *Encarsia formosa* has laid its eggs inside this whitefly pupa, resulting in the wasp pupa killing the whitefly. Beneficial insects can be an excellent way of controlling pests, especially in covered growing environments.
Above, right: Ladybirds are one of the most notorious consumers of aphids in the garden, although it is the larvae they attack most. They also eat pollen, so attract ladybirds to your garden with pollen-rich plants such as fennel (*Foeniculum*) or *Achillea*.

to stop them producing males. Balance is key. See also How to use biological controls, page 90.

It is harder to prevent diseases in a wildlife-friendly garden, but following Beth Chatto's mantra of 'Right Plant, Right Place' will mean that the plants you choose for your space are strong and healthy, and able to fight off pest and disease attacks. Good cultivation techniques, healthy, well-drained soil, and appropriate watering and feeding regimes will all encourage specimens that are able to hold their own when under attack. Resistant cultivars can be selected in some situations: for example, the elm (*Ulmus*) 'Wingham' has shown good resistance to Dutch elm disease (see page 103).

In writing this book, the horticultural experts at Kew have imparted some of their experience in sustainably managing pests and diseases. Reading these pages will help you to recognize certain symptoms and on what plants to expect to see them. The extensive plant collections at the Royal Botanic Gardens, Kew, mean that it has been impossible to mention every plant or issue, but we have collated the ones we come up against most frequently.

VEGETABLES AND FRUIT

BY HÉLÈNA DOVE

The kitchen garden is full of tasty plants, but sadly it is not just humans who want to consume these plants. Pests also enjoy the feast, while diseases are happy to make a home there. Most fruit and vegetable crops are highly cultivated, meaning they are more susceptible to pests and diseases. This chapter will help you to produce healthy crops in conjunction with nature.

Blight   *Phytophthora infestans* and *Alternaria solani*

PLANTS AFFECTED
Potatoes and tomatoes

DAMAGE CAUSED
Brown patches on foliage,
stems and fruits, leading
to rot and decay

Blight is a word that will fill a grower with dread as it is a
forever-looming disease that can wipe out entire potato and
outdoor tomato crops. Late blight (*Phytophtora infestans*) is an
oomycete, an organism similar to fungi but more commonly
called a water mould; it needs wet warm conditions to thrive
– 20–24°C (68–75°F). The first signs of attack are brown
patches on foliage with white, fungal-like haloes on the
undersides. The stems may show dark lesions, and the fruits of
tomatoes will begin to have dark spots. Once these signs appear,
it isn't long until the plant is doomed, as blight completes its life
cycle within five days, causing rot and a very recognizable smell.

If brown patches appear earlier in summer and have bullseye
concentric rings, this is early blight (*Alternaria solani*), which
will reduce the vigour of the plant, but the infected areas can
be removed and this should control the spread.

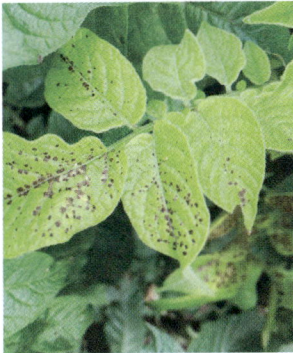

Top: Blight damage on plant stems.
Above: Symptoms of blight.

There is no sure-fire way to stop blight, but the risk can be reduced. Growing resistant cultivars can help with obtaining a harvest, although this tends only to slow the disease, not stop it and, as blight mutates fairly readily, new resistant varieties must be bred regularly. For tomatoes, cultivars such as 'Buffalosun' and 'Mountain Magic' can give a few extra weeks of cropping, and for potatoes 'Sarpo Mira' and 'Carolus' have shown to be resistant. Thankfully with early potatoes, blight tends to appear after harvest time.

Blight spores overwinter in the soil on volunteer potatoes – those that weren't harvested in the previous season; it's virtually impossible to find every last potato. Rotating crops is a great move if blight is present, as is mulching the soil to try and keep blight locked in and stop it spreading when water splashes from the ground to the leaves; also, always water at the base of the potatoes and tomatoes. Blight predominantly moves on the breeze, so spores can come from a fair distance to infect your plants.

Technology can help to tackle blight, as there are databases that monitor for the Hutton Criteria, and indicate when blight is likely. The Hutton Criteria is when an area has two or more days above 10°C (50°F) and six hours or more of 90 per cent relative humidity, so it's hot and wet weather. If it seems likely that blight is about to appear, the grower does have a chance of salvaging a crop. Tomatoes can be harvested, although sadly not stored. For potatoes, remove the haulms; there is then a chance that the tubers will remain unharmed in the ground for a few weeks. But keep checking and remove at the first signs of rot, which will show as reddish patches. All blight-infected material should go to the local waste, not be home composted, as the spores need an incredibly high temperature to be killed. If blight is a real problem, think about growing alternative crops, such as tomatillos instead of tomatoes and mashua for main-crop potatoes.

Flea beetles Various, including *Phyllotreta* and *Psylliodes*

PLANTS AFFECTED
Brassica family including
cabbages, radishes and rocket
DAMAGE CAUSED
Small holes in leaves

These small jumping beetles have a beautiful, blue or green metallic colour and are mainly active in spring and summer – not that you are likely to ever spot one as they jump away as soon as they are disturbed. Add to this that they are only 2–3mm/$\frac{1}{12}$–$\frac{1}{8}$in long, and it really is rare to see one unless purposely laying down a sticky trap to catch one.

The adults overwinter in leaf litter and appear in mid-spring, munching away on the leaves of most brassicas, such as broccoli, kale and swedes, leaving lots of small holes. They are particularly fond of young leaves on seedlings, and can cause a fair amount of damage and retard growth. In severe cases, the leaves may die off. Generally, however, the damage is aesthetic and most plants will grow through a flea beetle attack, although leafy crops such as rocket may be unsalvageable.

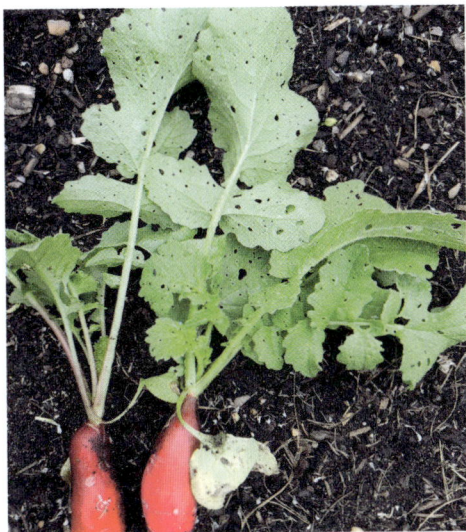

Above, left: Flea beetle damage.
Above, right: Adult flea beetles.

There are several ways to manage flea beetles in the garden. Growing crops under mesh netting will reduce their numbers; as most brassicas need to be grown under netting to ward off a host of other pests, this does make sense, although the amount of netting needed could make it a costly solution. Underplanting with radishes as a sacrificial catch crop is a fantastic way of managing flea beetles as well as utilizing the space; plant out large brassicas such as cabbage, sprouts and kale as large seedlings, and directly sow radishes between these. The flea beetles will meet the radish leaves before the larger brassica and eat this first. In best cases, the radishes will also grow to maturity and their roots will be left in a harvestable state.

Interplanting with pots of mint (*Mentha*), thyme (*Thymus*) or other strongly scented herbs also confuses flea beetles, protecting the brassicas (see How to use companion planting, page 52). Encourage wildlife such as frogs and ground beetles, which will feast on the larvae and so reduce the populations for the following year.

Codling moths *Cydia pomonella*

PLANTS AFFECTED
Apples and pears
DAMAGE CAUSED
Tunnels in the fruit bored
by the caterpillar

Codling moths are a pest in apple (*Malus*) and pear (*Pyrus*) trees, but generally will not deplete an entire year's harvest, instead affecting only some of the crop. Although in some years it does feel that codling moths have managed to ruin every fruit hanging on the trees, at least there are signs of the internal damage before biting into the fruit. The adult codling moths are fairly insignificant, having a wingspan of around 15mm/⅗in; they are a dull brown colour with bronze wing tips. These adults do very little damage, but will lay their eggs on the fruit and leaves of apple and pear trees in midsummer. The tiny, 1mm/½sin eggs contain the problem – the larvae – which, once hatched, bore into the developing fruits. Once inside, they seal up the entrance with apple bits, frass and a silk that they produce. They then bore away, making many tunnels until they mature and pop out of the fruit. By this point, the fruit is ruined. The damage internally causes the fruit to ripen prematurely and drop off – one of the signs of codling moth invasion. Another is the sink around the exit tunnel normally near the calyx, the opposite end to the stalk.

Below: Codling moth larva.

These caterpillars drop down into the leaf litter to pupate, and some will hatch several weeks later, to cause more damage, whereas most will wait until the next spring, when they will join the adult, egg-laying moths, ready to eat the young apples and pears. Thus, codling moths are active from late spring until late summer.

There are several methods to control codling moths, the first being to remove leaf litter and so reduce the chances of pupae finding anywhere to overwinter. Encourage birds into the garden; they will eat the caterpillars on their journey from leaf to fruit. Or else introduce pheromone traps. These work by emitting a synthetic version of the female sex-pheromone, enticing the males to the trap; once they enter the trap, they are caught by a sticky sheet and are unable to mate. These traps are designed mostly for monitoring, but can lower populations slightly as well. A combination of all these control methods should reduce codling moths or eradicate them altogether over several seasons, so the apples have a chance to ripen and be harvested by the gardener.

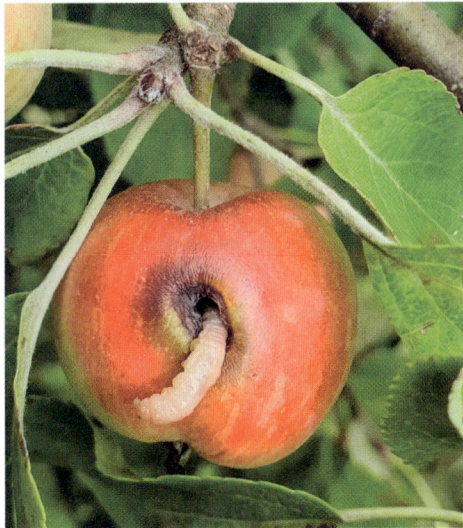

Chocolate spot *Botrytis fabae* and *B. cinerea*

PLANTS AFFECTED
Broad beans
DAMAGE CAUSED
Spots on leaves and pods

Top: Chocolate spot symptoms.
Above: Chocolate spot damage.

Chocolate spot is a delicious-sounding fungal disease that affects broad beans with its spores, which are spread in wet weather and temperatures of 15–20°C (59–68°F). It shows up as small brown spots on the leaves and the pods in late winter and spring; quite often, this is as far as the damage goes, meaning the resulting broad beans are still edible. However, plant growth is reduced. In more severe cases, when conditions are damper, the spots grow with concentric rings, causing the leaves to shrivel and die, and lesions may appear on the stems – all resulting in death of the plant.

To reduce the chances of chocolate spot, use seed from a reliable source, and never save seed from infected plants. When plants do show signs, remove the infected leaves and destroy. Increasing airflow will reduce any fungal infections, and rotating crops every year will also help minimize chocolate spot attacks. By planting in autumn and overwintering crops, the broad bean plants can often get to a good size and produce a harvest before being attacked by the fungus.

Clubroot *Plasmodiophora brassicae*

PLANTS AFFECTED
Brassica family including
cabbages, turnips
and wallflowers
(Erysimum × cheiri)
DAMAGE CAUSED
Swollen roots, stunted
growth and wilting

Above: Clubroot damage.

Clubroot is one of those incredibly frustrating diseases where the damage occurs underground and the only signs that anything is wrong show much later. The disease clubroot is a protist, and related to slime mould. Its spores enter brassica plants via root hairs, where it triggers cell multiplication, causing the root to swell beyond its normal size. Once the plant dies, spores spread into the soil and will persist for twenty years. Clubroot spores like wet and warm conditions (20–25°C/68–77°F) to spread so will normally attack from midsummer to early autumn.

Although it is extremely hard to grow brassicas in soil that has clubroot, and possibly the best advice is not to grow brassicas in such a place, there are some things that can be done to overcome this situation. Plant clubroot-resistant varieties; grow plants to a larger size than normal in their pots before planting out so there is more chance of a harvest; ensure soil has good drainage; and try to raise the pH slightly with products such as finely ground eggshells or seashells, because clubroot likes acidic conditions; add c.100g per sq. m/3½oz per sq. yard. One of these options alone probably won't overcome clubroot, but when combined they give the grower a fighting chance.

Cabbage root fly *Delia radicum*

PLANTS AFFECTED
Brassica family including cabbages, cauliflowers and swedes

DAMAGE CAUSED
Larvae eat the roots, eventually causing death of the plant

The adult cabbage root fly is about 6mm/¼in long and resembles a house fly and, as such, is relatively unnoticed by gardeners. The female lays her eggs on the roots of plants in the *Brassica* family, where the larvae hatch and feast on them. Once the damage to the roots has begun, it can be some time until the effects are seen above ground, and leaves start to wilt. By this point, the damage to the roots is too severe and the crop is lost.

The cabbage root fly usually goes through two life cycles in a year, but in warmer years can have three. The worst generation are those that hatch in early spring. The pupae from the late-summer generation overwinter just below the surface of the soil and hatch as flies in early spring. These adults mate and then lay their eggs in mid-spring, with the virtually transparent larvae hatching soon after. At this point, the majority of brassica crops have just been transplanted in the ground and have small root systems, which are destroyed by the larvae. Delaying planting until late spring can avoid the plants being the nursery for the eggs. The second generation of cabbage root flies hatch in late summer, when most brassica roots are larger and may be able to withstand a little damage.

The best way to protect crops from the cabbage root fly is with a barrier. There are two effective methods – mesh netting and root collars. Insect mesh netting must have a fine-enough weave to stop the 6mm/¼in fly entering the area. Such netting should be erected around the brassica bed; there is no need to add netting over the tops of the plants as the fly hunts low to the ground for somewhere to lay the eggs. Root collars can be made from any material, but cardboard is cheap and effective (see page 34). The root collar sits on the ground around the stem of the plant and stops the female from laying eggs on the roots.

Cabbage root fly larvae have some natural predators. Encouraging hedgehogs, birds and ground beetles to the vegetable plot will help control these pests. Ground beetles are encouraged by mulch, so add a layer, 5cm/2in deep, either before or after planting your brassicas.

How to make a cabbage collar

Cabbage collars can be used not only on cabbages but also on many brassicas such as Brussels sprouts, kale and calabrese. They protect the plants from cabbage root fly (see page 32), which will try to lay its eggs on their roots. As a secondary function, the collars also mulch the plants, stopping competition from weeds and keeping moisture in the ground. Cabbage collars can be purchased, and will last for several seasons, but making them is incredibly simple and a great job for wet days.

The cabbage collar needs to be a minimum of 15cm/6in diameter, but the larger the better: 30–40cm/12–16in diameter is ideal. It could be made out of any material but needs some flexibility to ensure complete contact with the soil. Thick fabrics such as old jumpers and tablecloths make great collars and may last for several seasons. Cardboard is the most common material used as it is available in abundance and will generally wear down by the end of the season, when it can be added directly to the compost bin as brown waste.

Check collars regularly and replace if needed. In a particularly wet season, cardboard collars may disintegrate to the point where a cabbage root fly can get herself in to lay eggs.

1. Choose a circular flower pot or plate of an appropriate size for your cabbage collar, and position it over your chosen collar material. Mark around the flower pot or plate, with a pencil. Cut out the marked area.

2. Cut a slit up to the approximate centre of the circle. This will be the entry way to get the plant stem into the collar. From this central point, cut three, small extra slits at right angles, then fold them upwards. This gives plenty of room for the plant stem to thicken. Plant your brassica, sinking it slightly lower than the soil line; this gives stability and also keeps the roots slightly lower, away from the cabbage root fly. Water well.

3. Add a root collar to each brassica plant by easing the stem through the long slit. To ensure contact with the soil, either put stones onto a cardboard collar or peg out a fabric one. Water the collar well, again ensuring contact with the soil, but also so stop the soil below drying out.

Large and small cabbage white butterfly

Pieris brassicae and *P. rapae*

PLANTS AFFECTED

Members of the *Brassica*
family, particularly cabbages

DAMAGE CAUSED

Holes in the leaves
and internal layers of
the cabbage heart

Cabbage white butterflies are notorious pests of the *Brassica* family. The butterflies of the large (*Pieris brassicae*) and the small cabbage white (*P. rapae*) are very similar in appearance, both having creamy white wings with black markings. The large cabbage white has a wingspan of 7cm/3in, and the males have black tips on their wings, while the females bear black spots on their forewings. The small cabbage white has one or two black dots and a wingspan of 5cm/2in.

Although the butterflies themselves do not cause the damage, they do lay their eggs on brassica leaves. The resulting caterpillars are the ones that destroy the crop, eating their way through the leaves, creating a Swiss-cheese effect. The large white caterpillars are yellow and black, and mainly devour the large outer leaves, whereas the small, green, barely noticeable small white caterpillars burrow into the hearts of cabbages, where they leave frass and destruction. Ultimately, such caterpillar damage means the crop is unharvestable.

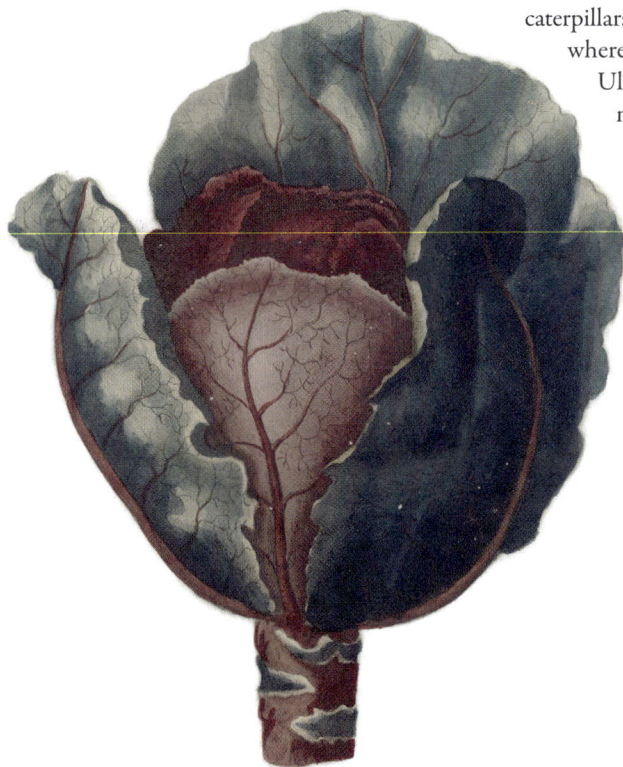

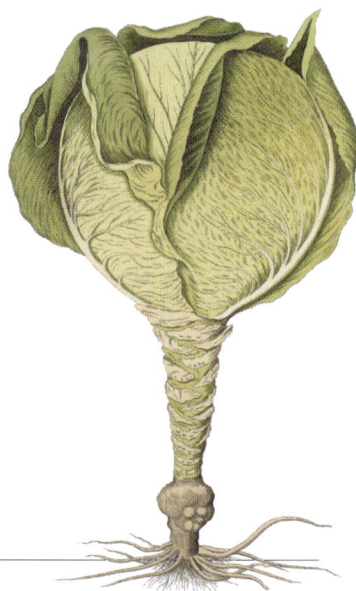

Above, left: Large cabbage white larva.
Above, right: Small cabbage white larva.

Barrier methods of protection remain the most effective way to keep the butterflies from laying their eggs on the plants. The cabbage white completes two to four life cycles a year, predominantly in late spring and late summer, but it is best to protect crops year-round. Netting the plants will always work best, ensuring that the holes in the mesh netting are around 6mm/¼in, to stop the small white squeezing through with its wings closed. The barrier also needs to be around a third taller and wider than the crops so that the leaves of the plants don't touch it, allowing the butterflies to lay their eggs through the holes. If there are still concerns, hand removal of eggs and caterpillars is very effective, as is a spray of the bacteria *Bacillus thuringiensis* (Bt) on the crops. Bt is a naturally occurring soil bacteria which, when consumed by the caterpillars, produces toxic crystals that kill the caterpillars. It does not have a harmful effect on other wildlife, but does need to be reapplied after rain showers.

Sacrificial plantings are also effective. Cabbage whites are attracted to nasturtiums (*Tropaeolum*), so underplanting the brassicas with this sprawling flower may give the caterpillars an alternate treat; as an added bonus, the leaves and flowers of nasturtiums are edible and taste delicious in a salad.

Cabbage whitefly *Aleyrodes proletella*

PLANTS AFFECTED
Brassica family including kale
and cabbages
DAMAGE CAUSED
Damage to outer leaves

Cabbage whitefly are tiny, sap-sucking insects that attack the outer layers of leafy brassica crops. They are found on the underside of the leaf, but will secrete honeydew, which in turn can cause sooty mould on the upper side. They generally don't cause a lot of damage and can be washed off once the crop is harvested. On plants such as kale where the leaves are the main harvest, insecticidal soaps can reduce populations (see How to treat your plants with insecticidal soaps, invigorators or oils, page 160). Meanwhile, ladybirds will feast on the cabbage whitefly, so encourage these beneficial insects.

Adult cabbage whitefly.

Asparagus beetles *Crioceris asparagi*

PLANTS AFFECTED
Asparagus
DAMAGE CAUSED
Leaves and stems

The asparagus beetle has a red head and spots on its back while the larvae are grey, and at all stages in the life cycle they feast on asparagus foliage in summer, leading to weaker harvests the following year. This pest also gnaws at asparagus stems, causing emerging spears to be deformed and any growth above this damage to go brown.

Hand remove the beetle, its larvae and eggs, which hang in lines off the stems, but be quiet and sneaky, as asparagus beetles often sense predators and drop into the soil. Encourage beneficial predators such as birds, frogs and ground beetles and apply the nematode *Steinernema carpocapsae*, which kills the larvae, in late spring and again when the second generation of asparagus beetles appear in late summer and early autumn.

Adult asparagus beetle.

Pea moths *Cydia nigricana*

PLANTS AFFECTED
Peas
DAMAGE CAUSED
Caterpillars eat the seeds
within the pod

Pea moth damage.

The pea moth lays its eggs on pea plants, with creamy coloured caterpillars hatching throughout summer. From the plant stem, they burrow into the pods and consume the juicy seeds, causing unseen devastation. Once each caterpillar has eaten its fill, it exits and pupates in the soil. Generally, it is very hard to notice that pea moth is present until the pod is harvested and opened.

Sow early or dwarf varieties such as 'Meteor' or 'Tom Thumb' to crop before midsummer, so avoiding the caterpillars entering the pods. Also, grow plants under mesh netting, to stop the moths laying eggs there. In milder areas, peas can be planted in mid-autumn and overwintered for an early harvest.

Pea and bean weevils *Sitona lineatus*

PLANTS AFFECTED
Peas, broad beans,
saved bean seed
DAMAGE CAUSED
Notches in leaves; damage
to saved seed

Pea and bean weevil damage.

Pea and bean weevils attack crops in spring and early summer, eating notches in the margins of the leaves. Thankfully, most plants can grow through this damage and it is just superficial. Because the damage is limited, there is little reason to control these weevils, but crop rotation can help. The weevils can smuggle their way within saved bean seeds, and eat the seeds when they hatch in spring. The best way to avoid this is to freeze any saved seed for a few days, to kill the eggs.

How to grow strong plants from seed

Although it is practically impossible to grow plants that will never be affected by pests and diseases, there are a few tricks to encourage resilience, especially when starting from seed.

AVOID DIRECT SOWING IF POSSIBLE

Seeds sown directly in the soil have many advantages and, of these, the seedlings that make it to maturity will often develop into incredibly strong plants, simply because they've had to outcompete the weeds, weather and pests and diseases thrown at them from a young age. But experience shows that not many plants will make it that far if the garden is full of slugs (see page 149), or if the ever-changing climate throws hot temperatures in early spring and hail in late spring. Thus, starting seeds in modules

or seed trays in protected areas gives each plant a good start in life.

When sowing in modules, resist the urge to pamper seedlings: don't feed at all – they usually don't need it and it is best to encourage them to put their roots out to find food. Thin out to give seedlings the space to build up a strong root network without competition.

HARDEN OFF

Hardening off is an important stage in growing strong plants as it prepares them for the harsh realities of open ground. This transition period from cosy house to turbulent garden starts with taking them outdoors during the day, and then covering with fleece or bringing them indoors overnight. After a week or so, leave the young plants open to the elements overnight,

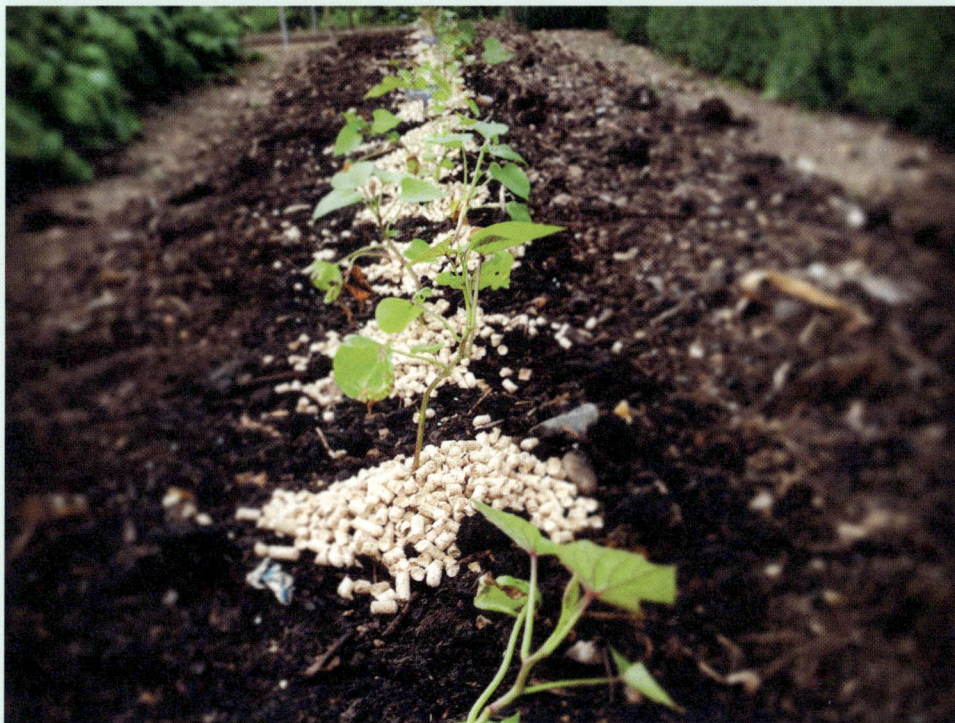

but in a sheltered spot. And then finally, once the soil conditions are favourable, transplant them into the ground, where they may sulk for a short period, but should then get away and put out their roots and reach for the skies.

TRANSPLANT BIG PLANTS

Avoid transplanting small plants as they are still incredibly susceptible to pests and competition from weeds. Growing plants to a large size helps in several ways: the leaves are further from the ground so less likely to be munched by slugs; and the leaves that are nearer the ground are the old ones, while the juicy new ones are further away from pests.

The more foliage a plant has, the more chance it has of regrowing if it does get nibbled, as it can photosynthesize more; plus, it will normally have a larger root system for seeking out the elements necessary for healing.

Opposite: Start plants, such as these lettuce, under cover in modules to give them the best start to putting out roots and growing strongly without predation.
Above: Transplant vegetable plants once they are large, growing strongly and fully hardened off. These sweet potatoes have been protected from slugs with wool pellets.

Corn smut *Mycosarcoma maydis*

PLANTS AFFECTED
Corn

DAMAGE CAUSED
Swollen kernels and blisters

Top: Symptoms of corn smut.
Above: Corn smut damage.

This fungal disease will mainly be seen in hot dry summers, usually of 27–33°C (80–91°F). It causes some of the kernels to swell up and become grey, which looks pretty gruesome. These swellings may eventually burst, spreading black spores everywhere. The spores may be carried on the wind for a fair distance and can persist in the soil for up to three years. In some cases, corn smut also causes blisters in all the plant parts.

There are several ways to manage corn smut, including buying resistant cultivars and rotating where the corn grows to hopefully avoid the spores that lurk in the soil. Another course of action is to the eat the swollen kernels before they burst. In Mexico, kernels infected with corn smut are thought of as a delicacy known as *huitlacoche* and fetch a high price at market, being said to have a mushroom flavour.

Common scab *Streptomyces scabiei*

PLANTS AFFECTED
Mainly potatoes, but also beetroot, carrots, swedes and turnips

DAMAGE CAUSED
Scabbing on the surface; possible death of tuber

Above: Potato scab symptoms.

There are two types of scab: powdery scab (*Spongospora subterranea*), which occurs in wet conditions; and common scab (*Streptomyces scabiei*), which develops when the ground is too wet and alkali. Both types of scab result in corky patches on tuber skins. This generally means the harvest is still edible, but, when scab is bad, the tuber skin can crack completely, and the tuber will rot. Common scab is most prevalent and is dealt with here.

Common scab is a bacterial disease that is related to slime mould and will attack the tuber in summer, entering through the lenticel or young skin; this means most first- and second-early potatoes are harvested before being affected. Sadly, the gardener won't know if their maincrop potatoes have been affected until they harvest the potatoes in late summer.

As common scab overwinters in the soil, rotate potatoes to reduce build-up, and always buy clean seed tubers. There are some resistant varieties such as 'King Edward' and 'Pentland Javelin', which give a greater chance of nice clean potatoes developing in the soil, as will ensuring there is a generous amount of organic matter in the soil to increase moisture levels. Therefore, mulch the area with 5–10cm/2–4in of compost or well-rotted manure in winter before planting, to encourage healthy soil.

Potato blackleg *Pectobacterium atrosepticum*

PLANTS AFFECTED
Potatoes

DAMAGE CAUSED
Stunted growth; yellow leaves; rotting of stems and tubers

Top: Potato blackleg damage.
Above: Symptoms of potato blackleg.

Potato blackleg is a bacterial disease that occurs in particularly wet seasons while temperatures are below 25°C (77°F). It is usually brought into the vegetable garden by infected seed potatoes, and can start to show its presence as early as late spring, when growth is stunted, with stems and leaves turning yellow. Eventually, if not removed, the stems will go black and rot, and any tubers produced will also rot.

The bacteria will overwinter on volunteer potatoes, so remove as many of them as possible, as soon as they sprout the next spring, and rotate where potatoes are grown to avoid bacterial build-up. Always buy certified seed potatoes, and immediately remove any plants starting to show signs of blackleg, to stop its spread. Blackleg is most likely to occur in wet conditions, so ensuring good drainage will help, as will choosing resistant varieties such as 'Charlotte'.

Parsnip canker *Itersonilia perplexans*

PLANTS AFFECTED
Parsnips
DAMAGE CAUSED
Cracked skin and brown-orange rot patches

Top: Symptoms of parsnip canker.
Above: Parsnip canker damage.

Parsnip canker is thought to be caused by several fungi, but primarily *Itersonilia perplexans*. It causes cracks in the skin and patches of brown-orange rot. If the canker is spotted, lift the roots; the non-affected areas of the parsnip can still be eaten. Should a cankerous parsnip be left for too long in the ground there will be little left to harvest for the winter table.

This disease thrives around 20°C (68°F) and in wet conditions, so generally attacks in autumn as the weather starts to cool and the rains start to ramp up. It is incredibly frustrating to lose mature plants.

There are several tactics to stopping parsnip canker, including ensuring good drainage in the planting site and using resistant cultivars such as 'Gladiator' and 'Archer'. Earthing up the rows of parsnips to cover their shoulders has been proven to help a little, as it breaks down the fungal spores and stops them attacking the parsnips. The spores can also travel on seed, so always buy from a reputable seed supplier.

Carrot fly *Psila rosae*

PLANTS AFFECTED
Carrots, parsnips,
celeriac and celery

DAMAGE CAUSED
Tunnels through roots

This almost invisible, little fly can cause a lot of damage to crops in the Apiaceae family, such as parsnips, celery and celeriac, but predominantly it haunts the carrot patch. The females pupate in spring and lay their eggs in late spring on the tops of the young carrot roots. The resulting maggots will crawl into the carrots and make a series of tunnels as they eat their way to maturity. If you are lucky these tunnels will be limited, so leaving some carrots to eat, but in some cases the tunnelling is too severe and the crop needs to be thrown away.

A second generation of carrot fly lays again, in late summer, although in some summers there is also a third generation, just to catch gardeners unaware. Not sowing carrot seed in the main egg-laying periods is one way of avoiding damage, although not foolproof.

Above: Carrot fly barrier.

Carrot flies locate their egg-laying site via scent, and when the leaves of carrots are disturbed they release a smell that the fly can find. Leaves are mostly disturbed during thinning seedlings, so sowing more thinly reduces how much scent is given out. However, carrot seed is notoriously hard to germinate, so don't sow too thinly. Masking the carrot scent with herbs such as chives (*Allium schoenoprasum*) and mints (*Mentha*) can confuse the fly; just remember to give the herbs a good rub after thinning carrots.

Sadly for the carrot fly, it is a poor flier, and can't go much higher than 60cm/24in, which is advantageous for the gardener. Thus, barrier methods such as mesh netting are always the best method of reducing pest attacks, and need to be in place before the crops are planted or sown. Carrots can be netted with micromesh, which stops the eggs being laid, but remember the fly is only 6mm/¼in long and can get through large netting. Erect mesh netting at least 70cm/28in high and preferably position it over the top of the carrot patch as well, as gusts of wind can help give fortuitous flies access to the carrots.

Resistant cultivars such as 'Resistafly' and 'Tozresis' are also available, and using this seed with a barrier, and maybe some strongly scented herbs dotted around the patch, is a good way of reducing the damage of this little fly.

Onion white rot *Stromatinia cepivora*

PLANTS AFFECTED
Onions, garlic, leeks
and chives
DAMAGE CAUSED
Discoloured foliage
and bulb rot

Top: Symptoms of onion
white rot.
Above: Onion white
rot damage.

Onion white rot is a soil-borne fungus that attacks members of the *Allium* family and is a complete nuisance. Its first signs are yellowing and wilting leaves above ground, then below-ground, white mycelial growth will rot the bulb. Within this white growth are small black units called sclerotia, which white rot produces instead of spores. These sclerotia can live in the soil for fifteen years with no host plant. This is partly what makes onion white rot so devastating. The best way to eradicate it is to not grow any *Allium* in the ground for twenty years, just to be on the safe side.

The sclerotia spring into life when they come into contact with chemicals that come from *Allium* roots. One potential way to get rid of the sclerotia is to trick it by using an onion soak: add water to pulverized onions, then leave to brew for a week before applying to the ground. This should force germination, with no host plant, but it may take several years of application to achieve a clean ground this way. Growing *Allium* in clean soil or compost in pots is the best way to achieve a harvest if the ground harbours this pesky fungus.

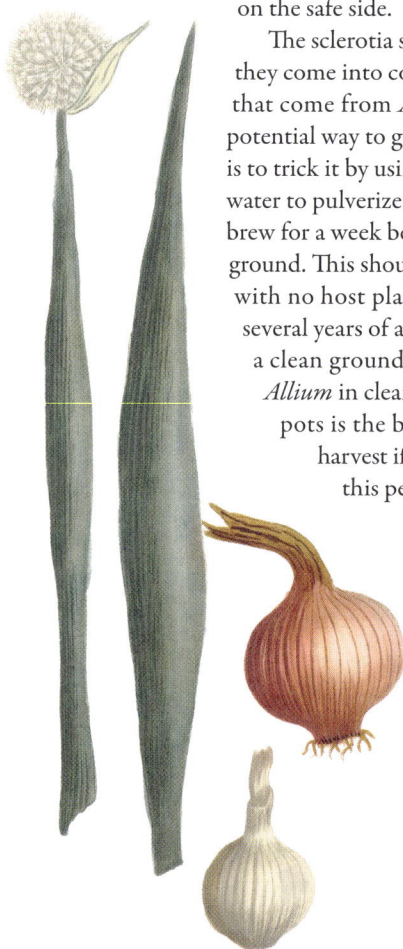

Leek moths *Acrolepiopsis assectella*

PLANTS AFFECTED
Leeks, onions and shallots

DAMAGE CAUSED
White patches; secondary rots

Leek moths are around 6mm/¼in long, with mottled brown wings. They hunt out members of the *Allium* family, such as leeks, shallots and onions, via scent, and lay their eggs in the leaves twice a year, in late spring and late summer. When the creamy coloured caterpillars hatch, they tunnel their way through the leaves, stems and bulbs of their host plants. These tunnels are seen as white patches by the gardener, but they also allow entry for secondary rots and can kill the plant. The best way to stop the moth laying is to grow the target crops under mesh netting.

Adult leek moth.

Fruit brown rot *Monilinia laxa* and *M. fructigena*

PLANTS AFFECTED
Apples, pears, plums and cherries

DAMAGE CAUSED
Rotten fruit; blossom wilt; spur death

Fruit brown rot is a fungal disease that is at its worst in warm humid conditions, around 25°C/77°F. The main symptoms are rotting fruits, which have brown pustules, from midsummer and often the fruit falling from the tree, although some will remain and mummify. This fungus can also cause blossom wilt and death to new shoots if it manages to get a foothold in early spring.

Although it is hard to irradicate, always ensure that each tree is pruned properly to give good airflow, and dispose of infected fruits and limbs to stop the spread.

Symptoms of fruit brown rot.

Strawberry leaf virus

PLANTS AFFECTED
Strawberries

DAMAGE CAUSED
Varied, but predominantly leaf deformation, stunted growth and low yield

Strawberries are subject to attack from many different diseases, including fungal, but the number of viruses has shot up in recent years. Many of the viruses display similar symptoms, and the way a gardener can treat them is very similar, so they are often clumped together. They can occur at any time in a strawberry plant's life cycle, throughout the year, but an infection becomes most obvious in midsummer.

Symptoms are initially evident on the leaves, but the visible issues are varied. They can include mottling, which can be seen clearly on the undersides of the leaves. Crinkling at the edges is becoming very common, as is the appearance of yellow spots or blotches, which sometimes turn red. The overall plant may become stunted. Occasionally, a plant will have one virus, but quite regularly it is attacked by several, which thoroughly weakens it. None of these viruses cause the plant to necessarily die, but it will lead to very low vigour and very few delicious strawberries for summer.

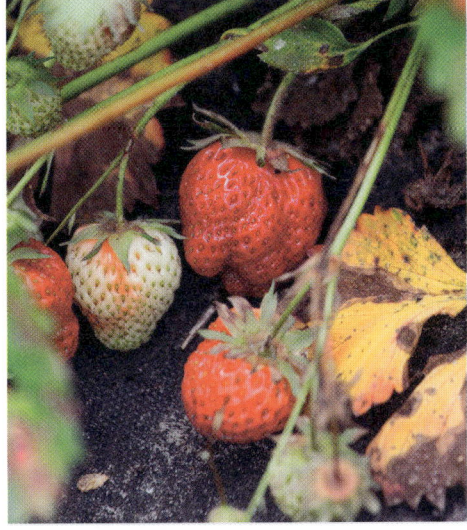

Above, left: Typical strawberry leaf virus damage to leaves.
Above, right: Typical strawberry leaf virus damage to fruit.

There are several ways to control viruses in strawberries, although it is hard to eradicate them completely. Always buy plants that are certified disease-free. These have often been micro-propagated, which is an amazing way to produce disease-free stock. Gardeners often enjoy collecting the runners from strawberries and passing them out to friends and family, but sadly these may contain viruses. Therefore, such practices need to be carefully monitored, so you don't pass on contaminated stock.

Viruses are spread via many vectors, one of the worst being aphids (see page 134). Reducing aphid attack will really help, so encourage beneficial insects such as ladybird and hoverfly larvae into the garden to control numbers (see page 19), and hand-squish any aphids on sight. Nematodes in the soil are also prone to spread strawberry leaf viruses, so don't replant fresh, virus-free stock where the last strawberry bed was, because the nematodes in this area are still likely to be carrying the viruses. Humans can also spread the viruses, and keeping tools clean really helps (see How to clean tools and other equipment, page 104). If working with infected plants, wash the blades of secateurs or other hand tools in a horticultural disinfectant to kill off anything, or even just wipe down the blades in between working on different plants.

How to use companion planting

It can often feel that pests are overwhelming in the garden, and the only solutions are ugly cages or plastic bird scarers. But companion planting offers a way of managing some pests while also creating a beautiful garden. There are two main forms of companion planting to help: sacrificial and beneficial.

SACRIFICIAL

Sacrificial plantings are those where, to protect a target plant, another is planted nearby to be consumed by the pest. A great sacrificial plant is nasturtium (*Tropaeolum*), which attracts aphids (see page 134) like no other, luring them away from other plants like beans and dahlias. It does mean the nasturtium will need to be pulled out after the attack, but allows the beans and dahlias to grow to a good size without being munched. Plant nasturtiums around the garden,

and ensure they have several leaves before the beans, dahlias or other susceptible plants are put in the ground.

Lettuces can be planted around any borders. As well as being quite beautiful, they will attract slugs (see page 149) very successfully, making them a fun sacrificial crop. It is down to the gardener to dispose of both the slugs and the spent lettuce, but this should reduce the numbers that reach precious other crops.

BENEFICIAL

Beneficial plantings can come in many forms, but usually either as a deterrent or else to attract beneficial insects for pollination and pest predation. Chives (*Allium schoenoprasum*) and other strong-smelling herbs such as mint (*Mentha*) are beneficial when used to deter

pests such as carrot fly (see page 46) that hunt by scent. Try growing such herbs in pots and sinking them into the soil where plants such as carrots and parsnips are growing. Give them a little stroke as you walk by, to release the scent, and move the pots around to where they are most needed.

Umbels such as false bishop's weed (*Ammi majus*) and ornamental carrot (*Daucus carota*)

attract insects such as hoverflies, whose larvae are some of the best at consuming aphids. These beneficial plants can be sown around borders, and their feathery foliage works well in most combinations. Start these plants early indoors to try and force flowering in spring, when aphids are most likely to attack young plants. Letting umbels self-seed generally gives a good early flowering – just try not to weed them out.

Far left: Polycultures mask crops from pests that hunt with scent or those that have weak eyesight. This sunflower (*Helianthus annuus*) hides the chard from birds flying overhead.
Left: Growing edible companion plants such as

pot marigolds (*Calendula officinalis*) attracts beneficial insects and provides a harvest, too.
Above: Ornamental carrots are full of pollen and great for attracting beneficial insects such as hoverflies and ladybirds.

Silver leaf *Chondrostereum purpureum*

PLANTS AFFECTED
Plums, cherries, apricots
and other members of
the *Prunus* genus

DAMAGE CAUSED
Silvering of leaves;
death of limbs

There is nothing better than a home-grown plum, cherry or apricot, but these stone fruits are prone to a fungal disease called silver leaf, which attacks when our backs are turned, in winter. The spores are released during the wet cool months, and enter healthy trees via wounds. The silver leaf fungus locates itself in the xylem, and slowly makes its way through the tree. In the following summer, some symptoms may start to be seen, although it can take a while for the damage to become evident, and by this point it can be too late to save the tree.

In all stone fruits, branches may bear silvery leaves, where toxins from the fungus have entered and separated the cells, giving the silver appearance. Once the leaves have turned silver, the branch itself will die off fairly soon after. Within the wood, a dark stain can be seen on all susceptible trees, which is a key indicator of silver leaf, which sadly means it can only be spotted when pruning. Another symptom is the appearance of bracket fungus on infected branches.

Although silver leaf is an incredibly destructive fungus, there are simple ways to stop its spread. Pruning susceptible plants in summer means that the wounds will have healed by the time the spores are active in autumn and winter. For this reason, it can be best to grow stone fruit as restricted forms such as fans, as these thrive with summer pruning. Remove branches as soon as possible if any symptoms appear. Cut back to healthy wood where feasible, to stop the spread, and clean tools well with a horticultural disinfectant (see How to clean tools and other equipment, page 104) when moving from branch to branch.

If planting a new plum tree where silver leaf is a concern, think about buying one on the rootstock Pixy, which has shown good resistance to the fungus. Pixy is also a dwarfing rootstock so lends itself to the restricted fruit-tree forms. Damsons and greengages show good resilience to silver leaf so could be an alternative stone fruit to consider.

Below, left: Typical silver leaf symptoms.
Below, right: Silver leaf damage.

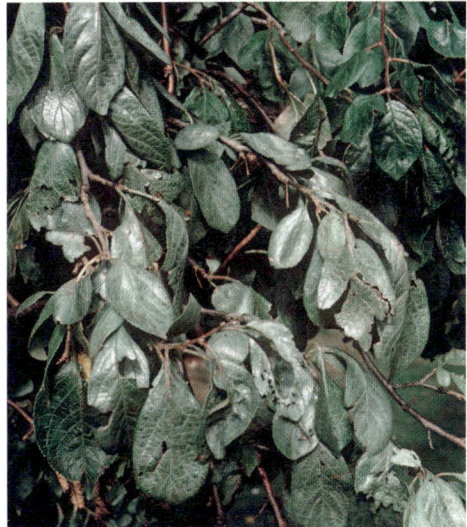

Blossom end rot

PLANTS AFFECTED
Large-fruited Solanaceae
plants such as tomatoes,
peppers and aubergines
DAMAGE CAUSED
Discoloration on the
bottoms of the fruits

Blossom end rot is a physiological disorder that appears primarily on container-grown tomatoes, but also on peppers and aubergines. Its symptoms are a dark sunken patch where the flower was, and it can spread to over half the fruit surface, with the insides becoming necrotic. The damaged area is then susceptible to secondary infections such as fungi and bacteria.

This disorder is related to a lack of calcium, and is usually caused through inconsistent access to water, not insufficient calcium in the soil. The plant takes up the calcium from the soil via its roots, and it travels around via the water vessels. When there is inadequate water, the limited supplies will go where it is pulled strongest via transpiration, which are the leaves, not the fruit, which will always be last in line. Calcium is an important part of the cell wall structure, and when there is a lack of calcium the cell membrane becomes less permeable, leading to the cell swelling and eventually cell death. Low amounts of calcium also reduce new cell formation.

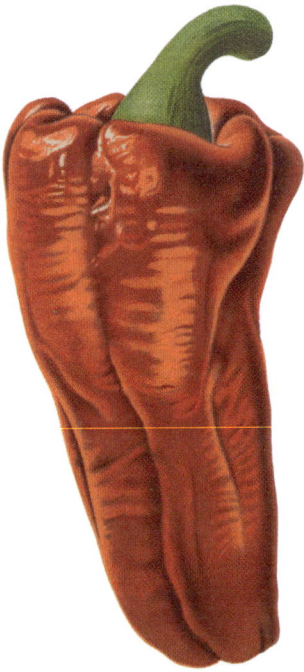

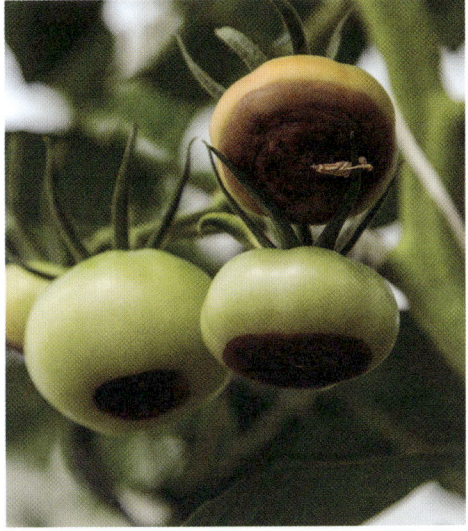

Above, left: Blossom end rot
on a pepper.
Above, right: Typical blossom
end rot damage on tomatoes.

Thankfully, blossom end rot is an easy disorder to correct
when caused by erratic watering, although the affected fruits
cannot be salvaged, so remove them. Simply ensure a consistent
supply of moisture in the soil. Remember to check below the
soil surface, usually 2–3cm/¾–1¼in down where the roots are;
if this feels moist to the touch, then there is no need to water,
but if it is dry then water the soil thoroughly. When growing
the plants in a pot, add water until it runs out of the base.
You should check several times a week, remembering that the
amount of water needed increases with growth, hot days and
fruit production.

If blossom end rot persists even with regular watering,
there may be another cause. Over-humidity in a greenhouse,
for example, stops transpiration and therefore movement of
water and calcium around the plant. This is easily solved by
ensuring good airflow around the greenhouse. Root damage or
insufficient growing space will also limit water flow, and once
again the poor fruits will be last in line for the calcium.

Apple and pear scab *Venturia inaequalis* and *V. pyrina*

PLANTS AFFECTED
Apples, pears, *Cotoneaster*,
firethorn (*Pyracantha*)
and *Sorbus*
DAMAGE CAUSED
Fruit, leaves and branches

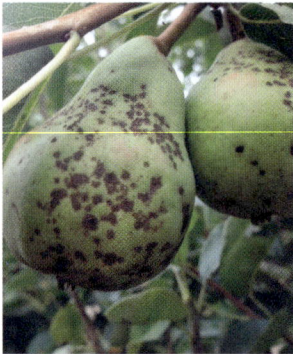

Top: Typical apple
scab damage.
Above: Typical pear
scab damage.

Apple and pear scab are two fungal diseases that wreak havoc in the orchard, from low-level annoyance to full-on fruit destruction, but thankfully for most home gardeners it's just an irritation. Although they are separate fungi, their symptoms and controls are almost identical, and both are much worse in a cool wet spring and summer. The scab spores overwinter in fallen leaves, and start to attack the plant from mid-spring. The scabs can appear on the leaves, as discoloured blotches, which cause early leaf fall; on twigs as blisters, which can crack and allow canker to enter; and on the fruit as brown scabs, which can crack causing the fruit to rot.

The best way to reduce scab is to collect and destroy all fallen leaves over winter. Prune out any infected branches and remove fruit with serious symptoms. There are some scab-resistant cultivars such as apple 'Discovery' and pear 'Beurré Hardy'.

Tomato brown rugose fruit virus *Tobamovirus* species

PLANTS AFFECTED
Tomatoes and sweet peppers
DAMAGE CAUSED
Discoloured fruits and plants

Typical tomato brown rugose
fruit virus damage.

Tomato brown rugose fruit virus (ToBRFV) is a relatively new virus, initially identified in Israel and first reported in the UK in 2019. At first glance, its symptoms are frustratingly similar to blight (see page 24), and include brown streaks on the stems and discoloured fruits with brown sunken patches, which lead to misshapen fruits. ToBRFV can be spread via seeds, but mainly through vectors, including humans, so ensure tools are cleaned between plants (see How to clean tools and other equipment, page 104) if suspected. As with other viruses, dispose of any fruits and plants showing signs of ToBRFV and always buy fresh, certified tomato seed.

Rhubarb leaf spot *Ramularia rhei* and *Ascochyta rhei*

PLANTS AFFECTED
Rhubarb
DAMAGE CAUSED
Discoloured leaves and stems

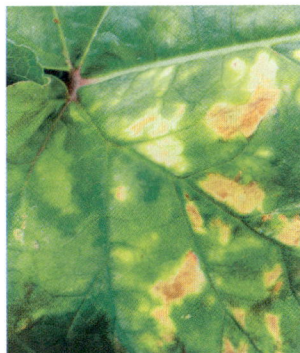

Typical rhubarb leaf
spot damage.

Often referred to as rhubarb rust, these two fungi attack the rhubarb in very similar ways. A common source is via kindly donated rootstock when someone is dividing up their rhubarb patch. The plant displays sunken red patches on leaves that spread to cover the entire area and generally cause it to die off early; also lesions may develop on the stems, which means less harvest for the gardener. The best way to avoid increasing this fungus is to carefully remove all stems and leaves, which the fungus overwinters in, and buy clean stock from a reputable nursery.

ORNAMENTALS

BY POLLY STEVENS

This chapter offers practical guidance on identifying, preventing and managing pests and diseases that commonly affect ornamental plants. By recognizing early signs and applying effective, environmentally conscious solutions, gardeners can protect the health of their plants, maintain their aesthetic appeal and ensure the vitality of their garden throughout the seasons.

Lily beetles *Lilioceris lilii*

PLANTS AFFECTED
Lilies (*Lilium*) and fritillaries
(*Fritillaria*)
DAMAGE CAUSED
Leaves, flowers and buds

These winged beetles are 8mm/⅜in long, with bright red bodies and black bellies, heads and legs. They can fly between plants, where they lay eggs on the undersides of leaves. These hatch into larvae, which are a reddish-brown colour and excrete a slimy black substance. Larvae feed on the leaves and then crawl into the soil, where they pupate. The adult beetles overwinter in the soil, hidden beneath leaf litter, where they are protected from frost and predators.

Signs of lily beetle presence can first be seen in spring, when adult beetles emerge from the soil and climb on to leaves, where they begin munching irregular holes in the foliage. The beetles can be picked off by hand; always place newspaper around the base of the plant first as they will quickly drop to the ground when disturbed. There are several natural predators of lily beetles, including frogs, birds and parasitic wasps (*Lemophagus errabundus*, *Tetrastichus setifer* and *Mesochorus lilioceriphilus*). Encouraging these predators and parasitoids into your garden will help keep the beetles at bay without the use of pesticides.

Top: Adult lily beetle.
Above: Damaged lily leaf with lily beetle larva and frass.

Chafer grubs *Phyllopertha horticola*

PLANTS AFFECTED
Most common on lawns
DAMAGE CAUSED
Bare patches on lawns; larvae
eat plant roots

Top: Adult chafer beetle.
Above: Chafer grubs.

The adults are brown winged beetles, 1cm/½in long, with a fine covering of hair. They emerge from pupae in late spring, feeding on nearby foliage and mating, then lay eggs among grass and other plants in early summer. The larvae take ten months to develop; during that period they live in the soil below plants while growing into large, cream-coloured grubs with orange heads, small front legs and black ends. Take care not to mis-identify stag beetle grubs.

Chafer grubs eat the roots of plants, weakening them and leading to plant death. This is a common problem on turf, where the grubs target grass roots; the grass then turns yellow, leaving large unattractive patches. Chafer grubs are a popular food for birds and mammals, which dig up the grass looking for them to eat, therefore causing more damage to turf.

Regular scarification of the grass will break down the thatch, bringing grubs to the surface and making them more readily available to birds. Using a heavy metal roller over a lawn, especially in early spring, will help kill pupae and emerging adults. Another effective method of reducing chafer numbers is by using the nematode and biological control *Heterorhabditis bacteriophora* applied during late summer or early autumn.

Pigeons *Columba livia*

PLANTS AFFECTED
Wide variety of ornamental
and edible plants
DAMAGE CAUSED
Leaves, flowers and seeds

Pigeons, also known as rock doves, are medium-sized birds with stout bodies, short necks and slender bills. They are typically 30–35cm/12–14in long, with a variety of plumage, the most common coloration includes shades of grey with iridescent green and purple feathers on the neck and chest. Their life cycle begins when a female lays one or two white eggs in a simple nest, often located in a sheltered spot. Both parents share incubation duties, which last 17–19 days. After hatching, the squabs (chicks) are fed with a nutrient-rich substance called 'pigeon milk' produced by the adults. Within 4–6 weeks, the young are fully fledged and ready to leave the nest. Pigeons can breed multiple times a year, making them prolific reproducers.

These birds primarily feed on seeds, grains, fruits and greens, although urban ones often scavenge scraps of human food. This adaptability allows them to thrive in various environments, from rural areas to bustling cityscapes. However, their feeding habits can cause significant damage to ornamental plants in gardens. Pigeons may peck at flower buds, eat seeds and strip leaves from young plants, affecting their growth and aesthetic appeal. Additionally, their droppings, which are highly acidic, can harm foliage and soil quality, potentially damaging the delicate ecosystem of a garden. Over time, their presence can lead to a decline in the health and vibrancy of ornamental plants.

Above: Adult feral pigeon.

Preventing and managing pigeons in your garden requires a combination of deterrents and habitat modifications. Physical barriers, such as wire netting, can protect plants from being accessed by these pigeons. Installing reflective objects like shiny tapes or wind chimes may scare them away, as they dislike sudden movements and bright reflections (see How to protect your plants from birds, rabbits and other mammals, page 74). Ultrasonic deterrent devices are also an effective option. Maintaining cleanliness by promptly removing food scraps and seeds from the ground will reduce the attractiveness of an area. Additionally, ensure that potential nesting sites, such as eaves or roof ledges, are blocked or modified to discourage pigeons from settling in or around the garden. By implementing these measures, gardeners can protect their ornamental plants and maintain a relatively pigeon-free space.

Box tree moths *Cydalima perspectalis*

PLANTS AFFECTED
Box (*Buxus*)
DAMAGE CAUSED
Defoliation

These moths are destructive pests native to East Asia that have spread to Europe and other regions such as North America. During their larval stage, they are typically bright green with black and white stripes running along their bodies, which are covered in sparse fine hairs. They grow up to 4cm/1½in long. The adult moths are white with brown edges or entirely brown, with a wingspan of 4cm/1½in. They lay clusters of pale yellow eggs on the undersides of box (*Buxus*) leaves, which hatch into caterpillars in a few days. These caterpillars feed exclusively on box plants, with a preference for the tender foliage and bark of young shoots. They strip leaves from the plant, leaving behind a webbed mass of silk, frass and damaged foliage. This feeding can cause extensive defoliation, weakening the plant and leaving it vulnerable to secondary infections and other stresses. In severe infestations, the caterpillars can kill entire box trees, significantly impacting their ornamental value. The damage is especially problematic in formal gardens or landscapes where box trees are used for hedges and topiary, as their precise shapes can be rapidly destroyed.

Box tree caterpillars go through several stages of development (instars) before pupating, often in cocoons hidden within webbing on the plant. Depending on the climate, there can be multiple generations in a year, making box tree caterpillars a persistent threat.

Below, left: Adult box tree moth.
Below, right: Typical box tree moth webbing.

It is essential to inspect box trees regularly for signs of eggs, caterpillars or webbing, especially during the growing season. Removing caterpillars by hand or pruning affected branches can help in small infestations. Alternatively, you can use *Steinernema carpocapsae*, a microscopic nematode available as a biological control to manage box tree caterpillar infestations. It works by entering the caterpillar's body through natural openings, releasing symbiotic bacteria that kill the host within forty-eight hours. To apply, mix the nematodes with water according to the manufacturer's instructions and spray directly on to box plants during cool overcast conditions, in early morning or evening to prevent desiccation of the nematodes. Ensure thorough coverage, especially on leaf undersides, where caterpillars hide. Keep the foliage moist for several hours after application, to allow nematode movement. Apply in spring or late summer, when caterpillars are small and active (see also How to use biological controls, page 90). Repeat treatments in severe infestations. Pheromone traps can monitor moth activity and help decide when to apply control measures.

Fuchsia gall mites *Aculops fuchsiae*

PLANTS AFFECTED
Fuchsia
DAMAGE CAUSED
Deformed leaves and
distorted flowers

Fuchsia gall mites feed by piercing plant tissues and sucking out cell contents, which causes significant damage. Their feeding disrupts the normal growth of fuchsia plants, leading to the development of abnormal galls, deformed leaves and distorted flowers. New shoots and flower buds are particularly vulnerable, as the mites prefer tender, actively growing tissues. The damage not only reduces the aesthetic appeal of the plant but also weakens its overall health. Severe infestations can stunt growth, prevent flowering and ultimately kill the plant if left untreated. The unsightly galls and deformations make the plant unattractive and unsuitable for ornamental purposes – a major concern for gardeners who prize fuchsias for their vibrant blooms.

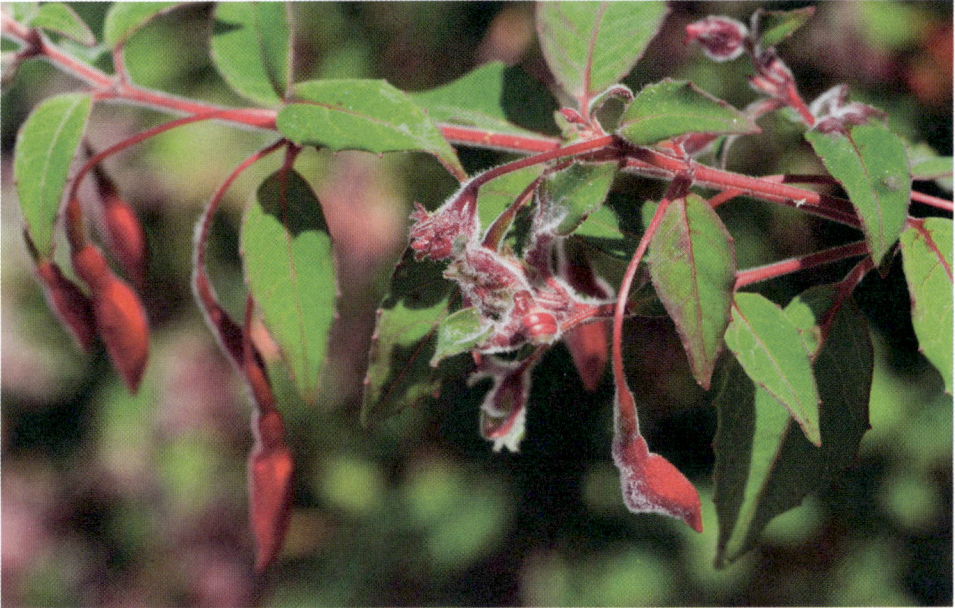

Above: Typical fuchsia gall mite damage.

Preventing and managing fuchsia gall mites requires early detection and proactive measures. Regularly inspect fuchsia plants for signs of galling or deformed growth, particularly in new shoots and flower buds. Prune and destroy affected parts immediately, to limit the spread of mites. Dispose of infected plant material in the general waste, as mites can easily transfer to healthy plants. Encouraging beneficial insects like predatory mites and lacewings can provide natural control (see page 19). Cultivating mite-resistant fuchsia varieties can also help to prevent infestations. Finally, maintaining overall plant health through consistent watering, feeding and pruning practices strengthens the plant's ability to recover from mite damage. By combining these strategies, gardeners can effectively manage fuchsia gall mites and protect their plants from severe harm.

Chrysanthemum leaf miners
Chromatomyia syngenesiae aka *Phytomyza syngenesiae*

PLANTS AFFECTED
Asteraceae family
DAMAGE CAUSED
Leaves

The larvae of this small fly species primarily feed on the leaves of *Chrysanthemum* and related plants, causing light brown, irregular lines that can affect both the health and aesthetic value of the plants. The adult flies are tiny, measuring 2–3mm/¹⁄₁₂–⅛in in length, with dark grey or black bodies and clear wings. Despite their small size, their impact on plant health can be considerable due to their reproductive habits and feeding behaviour.

The life cycle of the chrysanthemum leaf miner begins when the female lays eggs just beneath the surface of a leaf, typically near the edges. Within a few days, the eggs hatch, releasing small, pale yellow to white larvae. These larvae burrow into the leaf tissue, creating characteristic winding trails known as leaf mines. These mines appear as thin translucent paths on the leaf surface, where the larvae have consumed the inner mesophyll tissue. This feeding activity continues for 7–10 days, at which point the larvae mature and transition to the pupal stage. Pupation can occur either within the leaf or in the soil, depending on environmental conditions. Under optimal conditions, multiple generations of leaf miners may develop within a single growing season, exacerbating the problem and leading to an increased population over time.

The damage caused by chrysanthemum leaf miners can significantly weaken plants. As the larvae tunnel through the leaves, they disrupt the plant's ability to carry out photosynthesis effectively. This can result in a reduction in plant vigour, leading to leaf yellowing, browning and even premature leaf drop. In addition to affecting individual plants, infestations can spread quickly in greenhouse and nursery settings, where the controlled environment allows for rapid reproduction. This makes these leaf miners particularly problematic for gardeners who rely on chrysanthemums for their decorative appeal.

Above, left: Typical leaf damage caused by chrysanthemum leaf miners.
Above, right: Chrysanthemum leaf miner larva.

One of the first steps in control is regular monitoring. Gardeners should inspect plants frequently for signs of leaf miners and remove and destroy any affected leaves, to reduce the number of larvae present (see also How to catch pests with sticky friends, page 138). Use your fingers to crush the larvae within the leaf mines. Proper disposal of infested plant material is essential to prevent reinfestation, and can help break the pest's life cycle.

Encouraging natural predators is another effective method for managing these pests. Parasitic wasps, such as *Diglyphus isaea*, are known to target leaf miner larvae, helping to reduce their numbers. These beneficial insects can be introduced in greenhouses or encouraged in outdoor gardens through companion planting (see page 52) and habitat management. Floating row covers can also be used to protect plants from egg-laying adults, especially in outdoor settings. Additionally, maintaining healthy plants through relevant feeding, watering and pruning can improve their resilience to pest damage, making them less susceptible to severe infestations. A proactive and consistent approach will help preserve the health and beauty of chrysanthemums while reducing reliance on external interventions.

Rabbits Leporidae family

Grasses, leafy plants and wide
variety of ornamentals
DAMAGE CAUSED
Leaves/flowers chewed and
whole plants eaten

Above: Adult rabbit.

Rabbits are small mammals characterized by their long ears,
powerful hind legs and fluffy tails. They are prolific breeders;
females can produce several litters annually, each litter typically
containing 4–12 kits (baby rabbits). Rabbits thrive in diverse
environments, including grasslands, forests and urban areas,
although they're more commonly sighted in rural habitats.

These herbivores primarily eat grasses, leafy plants and
bark. Their feeding habits can cause significant damage to
ornamental plants, vegetables and shrubs. They often chew
on young shoots, flower buds and tender foliage, leaving plants
stripped and stunted. Rabbits also gnaw on the bark of young
trees, which can girdle and kill them. Additionally, rabbits dig
burrows for shelter, which may disturb soil and damage
plant roots.

Preventing and managing rabbits in a garden requires a
combination of exclusion, deterrence and habitat modification.
Physical barriers, such as wire netting, can protect individual
plants or garden beds. Deterrents such as motion-activated
sprinklers or ultrasonic devices can help scare rabbits away.
Reduce their hiding spots by
clearing brush, woodpiles or
dense vegetation to make the
garden less attractive to
rabbits. Also, encouraging
natural predators, such
as hawks, owls or
foxes, can help keep
rabbit populations in
check. See also How to
protect plants from
birds, rabbits and other
mammals, page 74.

Vine weevils *Otiorhynchus sulcatus*

Most ornamental and tuberous plants, especially pot-grown ones

DAMAGE CAUSED
Adults eat notches in plant leaves; larvae nibble at the roots, causing plants to wilt and eventually die

Top: Adult vine weevil.
Above: Vine weevil larvae.

The vine weevil is a flightless beetle commonly found in gardens and greenhouses. Adults are black/brown with a rough-textured body and a characteristic long snout. They measure around 10mm/⅜in long and are nocturnal feeders. The larvae are white, C-shaped grubs with brown heads, living in the soil and feeding on plant roots.

This pest has an unusual life cycle. Adult females reproduce parthenogenetically (without mating), laying up to 500 eggs annually in the soil. Eggs hatch into larvae that feed on roots through autumn and winter, and pupate in spring. Adult vine weevils emerge in late spring or summer, feeding on plant leaves and creating notched edges, while larvae attack roots, causing stunted growth, wilting and even plant death.

Prevention and management involve a combination of cultural and biological methods. Removing debris and applying sticky barriers (see How to catch pests with sticky friends, page 138) can limit adult movement. Regular plant monitoring helps detect early infestations. Using a biological control, such as the nematode *Steinernema kraussei*, will effectively target larvae in the soil (see page 90).

How to protect plants from birds, rabbits and other mammals

Protecting your plants requires a mixture of preventative measures, physical barriers and deterrents tailored to the type of intruder. Here are some practical strategies to safeguard your garden.

INSTALL PHYSICAL BARRIERS

Fencing is one of the most effective ways to keep mammals such as rabbits (see page 72) and deer out of your garden. Opt for wire netting with small gaps, as larger openings might allow smaller animals to squeeze through. For rabbits, ensure the fence is buried at least 30cm/12in underground to prevent these animals from burrowing under. Above ground, the fence should be 1m/3ft high for rabbits or 2m/7ft high for deer. For birds, netting draped over ornamental plants can effectively prevent pecking damage.

USE NATURAL DETERRENTS

Animals are often put off by strong smells or tastes they find unpleasant. Consider using organic sprays or placing strong scents like garlic or chilli pepper around the plants, to repel rabbits and deer. Crushed garlic and chilli can be mixed with water and sprayed on or around plants. A pre-mixed solution could also be left in a bowl beside plants to deter certain animals. Birds can be discouraged by reflective objects such as metallic streamers that create visual disturbances.

By combining these methods, you can protect your ornamental plants while maintaining a harmonious relationship with wildlife (see page 18). Regular monitoring and adjusting your strategies as needed will ensure your garden remains a beautiful thriving space.

Far left: Use plastic sheeting to cover seedlings.
Left: Mesh netting is excellent for protecting plants from birds and mammals.
Above: A bird scare humming line can be placed over seedlings. This makes a noise when catching the wind, which birds find off-putting.
Right: Ultrasonic devices are effective for scaring deer, rabbits and other animals. They contain a motion-sensor that emits high-pitched frequencies when activated.

Bulb rot in tulips

PLANTS AFFECTED
Tulips (*Tulipa*)
DAMAGE CAUSED
Bulbs develop brown/black
spots or become soft and mushy

This common problem affecting tulips (*Tulipa*) is caused by fungal or bacterial pathogens that thrive in moist or poorly drained soils. Fungal culprits include species such as *Fusarium, Botrytis* and *Pythium*, while bacterial soft rot is often caused by *Pectobacterium carotovorum*. Affected tulip bulbs typically display brown or black spots, soft or mushy textures and a foul odour in the case of bacterial infections. The disease can occur during storage, planting or growth, often spreading from one infected bulb to others. When planted, these bulbs may fail to sprout or produce weak stunted plants with yellowing foliage and poor blooms.

Tulip bulb rot symptoms include discoloration and decay of the bulb tissues, often starting at the base and spreading throughout the bulb. Infected bulbs may develop slimy or powdery textures, depending on the pathogen. If infected bulbs sprout, the resulting plants often exhibit yellow drooping leaves and underdeveloped flowers. The disease can severely impact ornamental displays, particularly in large-scale plantings, and can spread rapidly if infected bulbs are planted alongside healthy ones. Additionally, soil-borne pathogens can persist in the soil, infecting future plantings and causing recurring problems.

Preventing bulb rot in tulips begins with proper bulb selection and planting practices. Purchase bulbs from reputable suppliers to minimize the risk of infection. Always inspect bulbs before planting, discarding any that show signs of damage, discoloration or softness. To prevent soil-borne pathogens, plant tulip bulbs in well-drained soil. Avoid overwatering, as excessive moisture creates favourable conditions for pathogens. If soil drainage is a concern, amend the soil with sand or compost or plant the tulips in a raised bed or container.

Rotate tulip plantings annually to avoid depleting soil nutrients and to reduce pathogen build-up. During storage, keep bulbs in a cool, dry, well-ventilated area to prevent bacterial soft rot. If an infection occurs, remove and destroy affected bulbs and other plant material immediately, to prevent the spread of pathogens. By combining these preventative measures and timely interventions, gardeners can significantly reduce the risk of bulb rot and ensure healthy vibrant tulip displays.

Below, left: Typical bulb rot symptoms.
Below, right: Distorted flowers indicating a diseased plant.

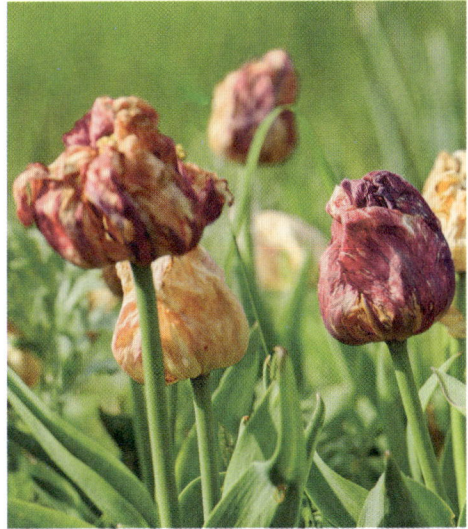

Fireblight *Erwinia amylovora*

PLANTS AFFECTED

Members of the Rosaceae
family that produce
pome fruits

DAMAGE CAUSED

Plant has a burnt appearance
with curled brown/black
leaves; blossom wilt;
dying shoots; oozing cankers

Fireblight is a highly destructive bacterial disease that primarily affects members of the Rosaceae family that produce pome fruits, such as apples, pears and quinces as well as ornamental plants like hawthorns (*Crataegus*), *Cotoneaster* and firethorn (*Pyracantha*). The disease gets its name from the scorched appearance it causes on infected plants. It spreads quickly in warm humid conditions, especially in spring and early summer, when plants are actively growing. The bacteria are often spread by rain, insects, wind or pruning tools, making it a persistent challenge for gardeners.

Symptoms typically start with water-soaked lesions on blossoms, stems or leaves that turn brown or black. The affected tissues often curl and remain attached to the plant, giving it a burnt appearance. Cankers may form on branches, exuding a sticky, amber-coloured liquid, which is a hallmark of the disease. As the infection progresses, branches may die back and, in severe cases, entire plants can succumb to fireblight. The disease not only reduces yield and quality on fruit trees but also significantly damages the aesthetic appeal and structural integrity of ornamental plants.

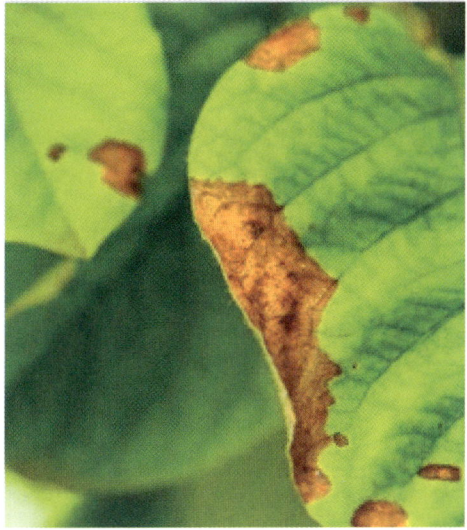

Above, left: Typical leaf and fruit damage from fireblight. *Above, right*: Early symptoms of fireblight.

Preventing fireblight requires an integrated approach focused on cultural practices, resistant plant varieties and prompt intervention. Start by selecting fireblight-resistant cultivars of apples, pears and ornamental plants whenever possible. Plant in areas with good air circulation to reduce humidity and avoid overcrowding, as this slows the spread of the disease. Proper sanitation is crucial when pruning infected branches; do this with clean tools during dry weather, making cuts 30–60cm/12–24in (depending on the size of the plant) below visible symptoms. Clean pruning tools between cuts with strong disinfectant or methylated spirit, to prevent bacterial spread (see page 104). Also, remove any secondary blossoms before they open.

Avoid using a fertilizer with high levels of nitrogen, which promotes the growth of succulent tissues that are highly susceptible to fireblight. Target irrigation at the base of plants, to minimize wetting the foliage and flowers. In the growing season, monitor plants regularly for early signs of infection and remove any affected tissues promptly. Severely infected plants may need to be destroyed to prevent further spread. By combining resistant varieties, good cultural practices and timely intervention, gardeners can effectively manage fireblight and protect their plants.

Azalea gall *Exobasidium japonicum*

PLANTS AFFECTED
Azaleas and other members
of the *Rhododendron* family
DAMAGE CAUSED
Leaves, flowers and shoots

Azalea gall is a common fungal disease that typically appears in spring, during cool and moist conditions, which favour fungal growth. The damage caused is primarily an aesthetic issue; however, severe infections can weaken plants over time if left unmanaged. The disease is most noticeable on evergreen azaleas, although deciduous varieties can also be affected.

The primary symptom of azalea gall is the formation of thickened, fleshy, pale green or white galls on leaves, flowers or shoots. These galls may grow to several times the normal size of the affected tissue and often feel spongy or waxy to touch. As the disease progresses, the galls may turn pink or white because of fungal spore development. Eventually, the galls dry out, turn brown and fall off.

While the disease does not typically kill the plant, it can weaken its overall health by diverting energy away from normal growth and reducing its aesthetic value.

Preventing azalea gall begins with good cultural practices. When planting azaleas, use well-drained soil with adequate air circulation; this will minimize moisture retention. Space plants appropriately to reduce humidity and promote airflow, which helps prevent fungal spores from taking hold. Avoid overhead watering, as wet foliage creates a conducive environment for the fungus.

Early detection is key to managing the disease effectively. Inspect azaleas regularly during the growing season, particularly in spring when galls are most likely to appear. If galls are found, remove promptly by picking them off by hand or by pruning the affected branches. Dispose of the infected material away from the garden, to prevent the spread of spores. Do not home-compost infected plant parts, as the fungus can survive and reinfect plants.

Fungicides are generally not used for managing azalea gall in private gardens. Maintaining overall plant health through proper fertilization, watering and mulching will enhance the plant's resilience to disease. By implementing these preventative and management practices, you can effectively control azalea gall and maintain healthy vibrant plants.

Rose black spot *Diplocarpon rosae*

PLANTS AFFECTED
Roses (*Rosa*)
DAMAGE CAUSED
Leaves

This common fungal disease is one of the most destructive for roses, affecting their health and aesthetic appeal. Rose black spot thrives in warm wet conditions and spreads rapidly during periods of rain or high humidity. It is especially problematic in gardens with poor air circulation or where plants are closely spaced.

The most noticeable symptom of rose black spot is the appearance of circular, black or dark brown marks on leaves. These spots are often surrounded by yellow halos; as the infection progresses, the affected leaves turn yellow and fall off the plant. The disease typically starts on the lower leaves and moves upwards as it advances. In severe cases, the defoliation can weaken the plant, reducing its flower production and ability to withstand other stresses, such as pests or drought. Rose black spot can also infect stems, causing small dark lesions, further compromising the plant's health.

Above: Typical rose black spot symptoms.

To combat rose black spot, start by selecting disease-resistant rose varieties, as these are less prone to infection. Recently developed rose cultivars should have a stronger resistance to disease than traditional ones. Plant roses in areas with good air circulation and plenty of sunlight, to minimize humidity around the foliage. Proper spacing between plants is crucial to reduce moisture retention. Mulching the base of the plants can help prevent fungal spores from splashing on to leaves during watering or rainfall. Water roses at the base rather than overhead to keep foliage dry.

Prune and remove infected leaves and canes promptly, disposing of them in the local waste rather than home composting, to prevent the spread of spores. Regularly clean up fallen leaves and debris around the base of the plant.

Maintaining overall plant health through suitable feeding and watering practices improves the rose's ability to resist infections and recover from damage. By combining these strategies, gardeners can effectively manage rose black spot and ensure their roses remain healthy and beautiful throughout the season.

Powdery mildew *Erysiphe, Podosphaera, Oïdium* and *Leveillula* species

PLANTS AFFECTED
Wide range of ornamental
and edible plants

DAMAGE CAUSED
White powder on plant
surfaces; stunted growth;
premature leaf drop

Typical powdery mildew symptoms.

Powdery mildew is a fungal disease that appears as a white powdery coating on leaves, stems and flowers. It thrives in warm humid conditions, spreading through wind-borne spores. The life cycle begins with spores landing on plant surfaces, germinating and forming fungal threads that produce more spores. Infected plants develop reduced photosynthesis, stunted growth and premature leaf drop. To prevent powdery mildew, plant disease-resistant varieties in areas with good air circulation, avoiding damp humid conditions. Prune overcrowded plants and remove infected foliage.

Replant disease

PLANTS AFFECTED
Wide variety of
ornamental plants

DAMAGE CAUSED
Poor growth; seedlings
or young plants may fail
to establish

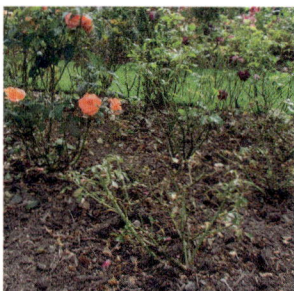

Typical replant disease symptoms.

Replant disease, or soil sickness, occurs when plants struggle to establish and grow in soil where the same or closely related species were previously cultivated (see How to establish young trees and shrubs, page 126). The condition is caused by a combination of factors, including the build-up of pathogens (for example, fungi, nematodes and bacteria), depletion of specific nutrients, and allelopathic compounds released by previous plants.

To manage replant disease, rotate crops with unrelated plant species. Try adding organic matter, replacing the soil or applying mycorrhizal fungi. Be sure to remove old roots and plant debris, to reduce the likelihood of replant disease.

Daffodil mosaic virus   aka *Narcissus* yellow stripe virus

PLANTS AFFECTED
Daffodils (*Narcissus*)
DAMAGE CAUSED
Leaves and flowers

Daffodil mosaic virus (DMV) is a plant virus that affects daffodils, causing yellow streaks or mosaic patterns. Infected plants may also experience stunted growth and distorted flowers, reducing their ornamental value. The virus spreads through infected bulbs, aphids (see page 134) and contaminated tools. It survives in bulbs and propagates systemically throughout the plant's tissues.

To manage DMV, use certified, virus-free bulbs and practise strict hygiene, including sterilizing tools and equipment (see How to clean tools and other equipment, page 104) and controlling aphid numbers. Remove and destroy infected plants immediately, to prevent further transmission, and rotate plantings regularly.

Typical *Narcissus* mosaic virus symptoms.

Fusarium patch   *Microdochium nivale*

PLANTS AFFECTED
Grass
DAMAGE CAUSED
Roots and blades of grass killed, leaving brown patches in a lawn

Fusarium patch, also known as snow mould, is a fungal disease affecting lawns. It appears as small, yellowish-brown patches that expand, with a slimy or white fungal growth visible in wet conditions. The disease thrives in cool damp weather, spreading through spores carried by water, wind or lawn equipment. Fusarium patch damages grass by killing roots and blades, leaving unsightly patches on the lawn. Prevention includes improving drainage, avoiding overwatering and aerating the soil to reduce compaction (see How to use low-tech methods in practical solutions for soil compaction, page 116).

Typical fusarium wilt symptoms.

Bulb eelworms *Ditylenchus dipsaci*

PLANTS AFFECTED

Bulbs (commonly daffodils/ *Narcissus* and tulips/ *Tulipa*), ornamental plants (such as *Phlox* and *Hydrangea*) and vegetables

DAMAGE CAUSED

Swollen distorted growth and eventual plant death

Bulb eelworms, also known as the stem and bulb nematodes, are microscopic nematodes that pose a significant threat to cultivated ornamental and edible plants. Measuring just 1–2mm/1⁄25–1⁄12in long, these eelworms are translucent and slender, making them almost invisible to the naked eye. They thrive in moist environments, moving through water films on plant surfaces or soil particles. These eelworms have a complex life cycle consisting of an egg, four larval stages and an adult stage, which can take as little as twenty days to complete under favourable conditions. Females lay eggs within plant tissues, ensuring a steady food supply for emerging larvae. These larvae can survive desiccation for extended periods by entering a dormant state, re-activating when moisture becomes available.

They feed by piercing plant tissue with their stylet (needle-like mouth part) and extracting cell contents. Their primary targets are bulbs, stems and leaves, where they cause extensive damage. Infested plants often exhibit symptoms such as swollen, distorted or stunted growth and brown necrotic patches on bulbs or stems.

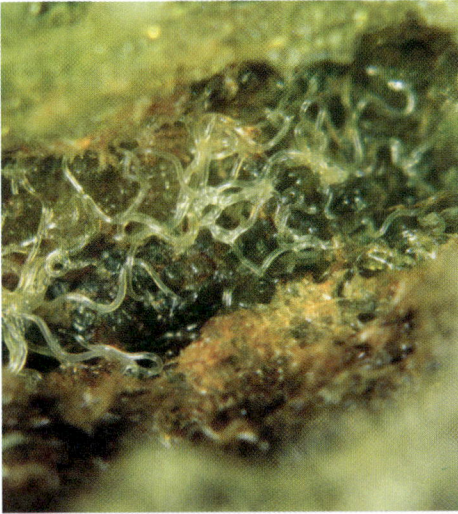

Over time, these nematodes disrupt nutrient flow within the plant, leading to wilting, reduced vigour and even plant death. Severe infestations can result in the complete failure of ornamental crops such as daffodils (*Narcissus*), tulips (*Tulipa*) and hyacinths (*Hyacinthus*). Additionally, bulb eelworms spread easily through contaminated soil, water or infected plant material, making them a persistent and challenging pest to manage.

Preventing bulb eelworm infestations requires a combination of good cultural practices and vigilant monitoring. Start by using certified, disease-free bulbs and planting material, to minimize the risk of introducing eelworms. Crop rotation with non-host plants can help reduce their populations in the soil, so try to avoid planting bulbs in the same area for three years. Maintaining well-drained soil and reducing irrigation also create less favourable conditions for the eelworms.

For infested areas, heat treatment of bulbs – soaking them in water at 43–45°C/109–113°F for about three hours – can kill the eelworms without damaging the plants. Additionally, removing and destroying infected plant material helps prevent the spread of the pest. By combining these strategies, growers can effectively manage eelworm damage and safeguard their ornamental plants.

Solomon's seal sawfly *Phymatocera aterrima*

PLANTS AFFECTED
Solomon's seal (*Polygonatum*)
and other ornamental plants
DAMAGE CAUSED
Small holes/irregular
notches in leaves and
complete defoliation

The adult Solomon's seal sawfly is a small black insect measuring 6–8mm/¼–⅜in in length. It resembles a tiny wasp, but lacks the narrow waist typical of a true wasp. The sawfly larvae are the most damaging for a plant. They are pale greenish white with black heads, and grow to 18mm/¾in in length – they are often mistaken for caterpillars due to their soft elongated bodies. The larvae may also have a slightly translucent appearance, making them blend with the foliage.

Solomon's seal sawfly has one generation per year, with its life cycle closely tied to the growing season of its host plants. Adult sawflies emerge in late spring or early summer, often coinciding with the emergence of Solomon's seal foliage. Females lay eggs on the undersides of leaves, typically along the midrib. After a few days, the eggs hatch into larvae, which begin feeding voraciously on the leaves.

The larvae pass through multiple growth stages (instars) and continue feeding for several weeks, skeletonizing or completely consuming the foliage. Once fully grown, the larvae drop to the soil, where they pupate. The pupal stage lasts through autumn and winter, with adults emerging during the following year to repeat the cycle.

Above, left: Sawfly larva.
Above, right: Typical Solomon's seal sawfly damage.

The primary damage is caused by the feeding activity of the larvae. Early signs of infestation include small holes or irregular notches in leaves. As the larvae grow, they can strip entire leaves, leaving only the midribs. Severe infestations may completely defoliate the plant, weakening it and making it more susceptible to other pests and diseases. While Solomon's seal plants are perennial and may survive defoliation, repeated infestations over several years can significantly reduce their vigour and ornamental value.

To control these pests, regularly inspect Solomon's seal plants, especially in late spring and early summer, for signs of eggs or larvae. Early detection allows for prompt intervention before significant damage occurs. Manually remove larvae from leaves by hand-picking them or shaking the plant to dislodge them. Destroy the larvae to prevent them from pupating and continuing their life cycles.

Maintain plant health by ensuring proper watering, mulching and feeding. Remove plant debris in autumn, to help reduce overwintering sites for pupae. Encourage natural predators such as birds, ground beetles and parasitic wasps (see Wildlife-friendly solutions, page 18), which may help control sawfly populations. By combining these strategies, you can effectively manage Solomon's seal sawfly infestations and protect your ornamental plants.

How to use biological controls

When it comes to managing pests and diseases with biocontrols, you need to understand your plants and the growing environment. It's essential to know which pests your plants are vulnerable to (see page 14), and what time of year is best to take action and prevent pest outbreaks from ruining your garden.

MONITOR YOUR PLANTS FOR PESTS REGULARLY

When a pest is spotted early, dealing with it becomes much easier, and this is particularly important when adopting a biological approach. If you release a high number of beneficial insects to a relatively low population of a pest you will gain control quickly and effectively. If, however, you have a large pest outbreak you either need to wait a long time for the beneficials to increase in numbers or else need to release a lot more biologicals. In most cases like this, it is best to reduce the pest population first before employing a biological control. This could be by pruning out infestations, manual removal or using an insecticidal spray (see page 160).

UNDERSTANDING YOUR PEST IS ALSO KEY

It is important to ask questions such as: what is the pest, and what is the pest's life cycle? Knowing what pest you are dealing with can help you choose the best treatment, and understanding its life cycle will allow you to release the relevant biological at the best time. Consider all the pests infecting your plants. Dealing with one pest effectively with biological controls, yet not managing another, can be challenging especially when deciding if you should use insecticides to tackle the problem, as these can inhibit biocontrols.

BIOLOGICAL MEASURES

Effective biological control starts by organizing a schedule of release times for preventative measures, where to release each treatment and

for what purpose. At the beginning of the growing season, apply your chosen beneficials. Place them on the plants or as close as possible if supplied in containers. This could be on the stems and leaves or surface of the soil. Then monitor both the pests and the biologicals. Top up your biologicals through the year if needed.

MICROBIAL CONTROLS

Microorganisms such as bacteria, fungi and nematodes can also effectively target specific pests. Such biological agents are available in liquid or powder form and should be applied according to the manufacturer's instructions.

Left: Biological controls can be applied with a watering can or a sprayer.
Above: Work systematically over the affected area when using a biocontrol.

Below are examples of microbial controls:

- *Steinernema carpocapsae*: A naturally occurring nematode that has evolved a symbiosos with bacteria that produces toxins harmful to some caterpillars and beetles – it has been developed to target only specific pests. It is good for control of box tree caterpillar (see page 66) and can be applied to box (*Buxus*) using a sprayer.
- *Heterorhabditis bacteriophora:* A beneficial nematode which attacks soil-dwelling pests like grubs and root maggots, and can be sprayed over infected lawns. *Heterorhabditis* will search for chafer grubs (see page 63) in the soil and enter their bodies, feeding off the grubs and reproducing, eventually killing the host chafer grub.

Sclerotinia *Sclerotinia sclerotiorum*

PLANTS AFFECTED
Dahlia and other
ornamental plants
DAMAGE CAUSED
Wilt, rot and plant death

Sclerotinia, also known as white mould, is a fungal disease that leads to wilt, rot and plant death. The disease is characterized by white, cottony fungal growth on stems, leaves and flowers (often affecting *Dahlia*). Hard black resting structures called sclerotia form within infected tissue or soil, enabling the fungus to survive adverse conditions.

The life cycle begins when sclerotia germinate in moist cool conditions, releasing airborne spores that infect plants. The fungus enters through wounds or directly invades plant tissues, causing wilting, browning and rotting. Infected stems may collapse, and fluffy white fungal growth becomes visible. Sclerotinia thrives in damp environments, often spreading through splashing water, contaminated tools or soil. Damage includes rapid decline and death of affected plants, reducing their ornamental value.

To prevent sclerotinia, ensure proper air circulation, avoid overwatering and space plants to minimize humidity. Remove and destroy infected plant parts promptly. Crop rotation and keeping the garden clean of debris help reduce fungal persistence. Use resistant plant varieties and maintain overall plant health to reduce susceptibility to sclerotinia further.

Top: Typical sclerotinia
stem damage.
Above: Sclerotinia
flower damage.

Phytoplasmas *Candidatus phytoplasma*

PLANTS AFFECTED
Wide range of ornamental plants including *Aster, Dahlia* and roses (*Rosa*)

DAMAGE CAUSED
Yellowing stunted growth; abnormal leaf development; proliferation of shoots; flowers may be deformed, discoloured or fail to bloom entirely

Above: Typical phytoplasma damage.

These microscopic, wall-less bacteria infect plant phloem tissue, disrupting nutrient transport. They are responsible for diseases like aster yellows and witches' broom, affecting many ornamental plants. Phytoplasma infections weaken plants, reduce their ornamental value and can eventually lead to death.

Phytoplasmas are transmitted by sap-feeding insects, primarily leafhoppers (Cicadellidae family), which acquire the pathogen while feeding on infected plants. Once inside a plant, phytoplasmas multiply and spread through the phloem, affecting all parts of the plant. They overwinter in perennial host plants or insect vectors, continuing the disease cycle in subsequent growing seasons.

Prevention and management focus on reducing the presence of insect vectors and removing infected plants. You can control leafhopper populations with physical barriers like mesh netting. Maintaining garden health through proper watering, feeding and pruning reduces stress, making plants less susceptible to bacteria. Destroy infected plants to prevent further spread, and choose resistant varieties when buying plants.

TREES

BY KEVIN MARTIN

This chapter is designed to help you keep your trees healthy, resilient and looking their best. Alongside practical guidance on some of the most common tree pests and diseases you may encounter in your garden, you will find useful tips for caring for your trees – from planting the next generation and maintaining clean pruning tools to addressing the often-overlooked issue of soil compaction.

European gypsy moths *Lymantria dispar*

PLANTS AFFECTED
Broadleaved trees and shrubs
DAMAGE CAUSED
Leaves

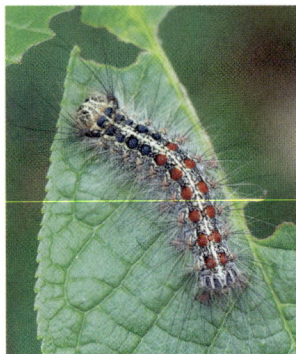

Top: Adult European
gypsy moth.
Above: European gypsy
moth caterpillar.

The European gypsy moth is widely distributed throughout most of London and south-eastern England. It is also native to parts of continental Europe, where populations periodically reach very high numbers. Its caterpillars feed on the foliage of various broadleaved trees and shrubs, and have a preference for oaks (*Quercus*) and poplars (*Populus*) in forests. In gardens, the gypsy moth also targets small trees and shrubs, including ornamental conifers, as well as hedges, such as those of beech (*Fagus*). High population densities can lead to severe damage, especially on small trees. While most plants recover without losing vigour, repeated infestations can weaken trees, compromise their health and eventually cause death.

To manage gypsy moth populations, a combination of methods is effective. Natural substances like azadirachtin, derived from neem tree seeds, can kill larvae. Pheromone traps help monitor populations and disrupt mating by preventing males from locating females. Wrapping fabric or duct tape around tree trunks traps caterpillars moving up and down the trees. These combined measures can protect trees and shrubs from severe damage. However, if the problem continues, try spraying pesticides at dusk, when caterpillars are active; this works well, particularly on younger larvae, which are more susceptible to insecticides.

Grey squirrels *Sciurus carolinensis*

PLANTS AFFECTED
Common beech (*Fagus sylvatica*) and sycamores (*Acer pseudoplatanus*)
DAMAGE CAUSED
Bark

Above: Adult grey squirrel.

Since their introduction into Britain between 1876 and the 1920s, grey squirrels have spread rapidly, displacing the native red squirrel (*Sciurus vulgaris*) through competition for food and transmission of the squirrel pox virus. Grey squirrels are now found in most of England and Wales, as well as central and south-eastern Scotland. They are highly destructive in woodlands, stripping bark from tree stems and branches. Common beech (*Fagus sylvatica*) and sycamore (*Acer pseudoplatanus*) trees are often the most severely affected, but damage also occurs on oaks (*Quercus*), birches (*Betula*), larches (*Larix*), pines (*Pinus*) and Norway spruces (*Picea abies*). Bark wounds can cause deformation, staining and decay of timber, and severe injuries may result in tree death. This level of destruction has become a major deterrent for landowners seeking to establish new woodlands.

With grey squirrels being widespread throughout the UK, it is extremely difficult to control them using traditional methods such as trapping. However, an innovative approach is under development as part of the UK Squirrel Accord's (UKSA) grey squirrel fertility control programme. This research project utilizes feed hoppers to administer an oral contraceptive, aiming to reduce population growth in an effective and humane way. Such measures could help mitigate the significant environmental and economic impact caused by grey squirrels.

Phytophthora root rot *Phytophthora*

PLANTS AFFECTED

Maples (*Acer*), horse chestnuts (*Aesculus hippocastanum*), *Aucuba*, box (*Buxus*), California lilacs (*Ceanothus*), Lawson's cypresses (*Chamaecyparis lawsoniana*) especially 'Elwoodii', holly (*Ilex*), lavenders (*Lavandula*), apples (*Malus*), cherries (*Prunus*), *Rhododendron*, flowering currants (*Ribes*), raspberries (*Rubus idaeus*), *Sorbus*, yews (*Taxus*, notably very susceptible) and *Viburnum* among others

DAMAGE CAUSED

Root and stem-base decay

Phytophthora root rot is second only to honey fungus (see page 100) as a leading cause of root and stem-base decay in trees and shrubs. Numerous *Phytophthora* species cause similar symptoms, affecting woody plants, herbaceous perennials, bedding plants, pot plants and bulbs. Environmental changes, such as milder winters, wetter springs and hotter drier summers, exacerbate the problem. Increased rainfall promotes pathogen survival and spread, while drought stress weakens trees, making them more susceptible to infection.

These pathogens persist over winter in infected hosts, soil and plant debris as dormant spores. In water films, spores swim and spread via rain, watercourses, and run-off. Some species, such as *P. ramorum*, disperse aerially through mist and wind-driven rain. Human activity, including the movement of infected soil, plants or equipment, further facilitates transmission. *Phytophthora* is particularly damaging as it attacks roots, collars and lower stems, causing cankers and lesions. Fine feeder roots rot away, hindering water and nutrient uptake, leading to wilting, sparse or yellowing foliage, and branch dieback. In conifers, needle discoloration progresses from green to brown.

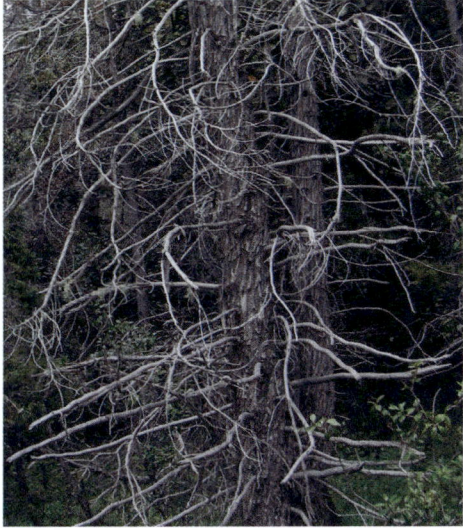

Above, left: Phytophthora root rot symptoms.
Above, right: Typical damage from phytophthora root rot.

Below-ground inspections reveal fine root decay, while larger roots may develop soft brittle interiors. In severe cases, infection spreads to the collar or stem base, causing weeping lesions or bark discoloration. However, similar symptoms can result from waterlogged soils or other root diseases, making laboratory testing essential for accurate diagnosis.

While foliar *Phytophthora* diseases, such as late blight (*P. infestans*; see page 24), are well known to gardeners, root rot presents unique challenges. Effective management focuses on improving soil drainage, minimizing waterlogging and preventing contamination through soil or plant movement. Early detection is crucial. In newly affected areas, infected plants should be promptly removed, and the surrounding soil replaced with fresh topsoil. Replanting with less susceptible species reduces re-infection risks.

Certain plants are highly vulnerable to *Phytophthora* in the UK and should be avoided in areas where the disease is prevalent. Choosing resistant alternatives and maintaining optimal soil conditions can help curb the spread of this destructive pathogen.

Honey fungus *Armillaria* species

PLANTS AFFECTED
Woody and herbaceous
perennials

DAMAGE CAUSED
Root system failure

Honey fungus is the common name for several species of fungi
within the genus *Armillaria*. This fungus spreads underground,
attacking and killing the roots of plants, and then decaying
the dead wood. It is the most destructive fungal disease in
UK gardens and can attack many woody and herbaceous
perennials; unfortunately, no plants are completely immune.
Symptoms include the sudden death of the upper parts of the
plant, especially during hot dry weather, indicating root system
failure. In some cases, the death occurs more gradually, with
branches dying back over several years. Other signs include
smaller, paler-than-average leaves, failure to flower or unusually
heavy flowering followed by an excessive crop of fruit, usually
just before the plant's death. Visible bleeding from the bark,
particularly around the base, and dead decaying
roots, with sheets of white fungal material
(*mycelium*) between bark and wood,
can also be observed. The smell of
mushrooms often accompanies this
decay. Rhizomorphs, which look
like black bootlaces may also be
present in the soil, although they
are difficult to detect.

Above, left: Honey
fungus toadstools.
Above, right: Honey fungus
growing on an old tree trunk,
where rhizomorphs and
whitish mycelium sheets
are also visible.

Unfortunately, no chemicals are currently available to kill
honey fungus. However, one method of control is to remove
any dead or decaying woody material from the affected area,
eliminating a food source for the rhizomorphs and slowing
their growth. To prevent the spread of honey fungus to other
uninfected areas, a physical barrier, such as a vertical strip,
45cm/18in deep, of butyl rubber (pond lining) or heavy-duty
plastic sheet, can be buried in the soil. This should protrude
about 2.5cm/1in above soil level. Regular deep cultivation
can also help, by breaking up rhizomorphs and limiting
their spread.

Selecting plants well suited to the growing conditions is
another way to manage honey fungus. Plants that are stressed
by environmental factors, such as drought or waterlogging, are
particularly vulnerable to attack. By choosing those that are
more resilient, you can reduce plant stress and make them less
susceptible to honey fungus.

Sooty bark disease   *Cryptostroma corticale*

PLANTS AFFECTED
Maples (*Acer*), especially
sycamores (*A. pseudoplatanus*)
DAMAGE CAUSED
Dieback and death

Above, top: Flaking bark on
dead sycamore tree.
Above, bottom: Typical sooty
bark disease damage.

Sooty bark disease, first described in North America, was
introduced to Europe with the first UK case documented in
1945. It has since spread further across central and western
Europe, particularly since the 2000s. Its impact is linked to
higher summer temperatures and drought, raising concerns
with recent hot dry summers and future climate change. Spores
from sooty bark disease also pose human health risks. Although
rare, exposure to large quantities of spores can cause maple bark
stripper's disease, a type of hypersensitivity pneumonitis.

The fungus infects plants through fresh wounds, leading to
wilting within weeks. The mycelium progressively colonizes the
wood, spreading rapidly in sapwood but more slowly in bark.
Stress from heat and drought can trigger sooty bark disease
symptoms as the fungus invades the outer wood layers, kills the
cambium and produces stromata (fungal fruiting bodies). These
release dark spores when the bark splits.

Control options are limited to removing affected tree
parts or infected trees entirely. Preventative measures include
maintaining healthy trees, such as by providing irrigation
during drought, to reduce stress and vulnerability to infection.

Cypress aphids *Cinara cupressivora*

PLANTS AFFECTED
Hedging conifers such
as Leyland cypresses
(× *Hesperotropsis leylandii,*
aka × *Cuprocyparis leylandii*),
Monterey cypresses
(*Hesperocyparis macrocarpa,*
aka *Cupressus macrocarpa*)
and Lawson's cypresses
(*Chamaecyparis lawsoniana*)

DAMAGE CAUSED
Yellowing shoots turning
brown, with severe dieback
on clipped hedges

Cypress aphids are blackish-brown, sap-sucking insects that
feed on the stems of hedging conifers, causing foliage to
brown and dry out, particularly in summer. A black sooty
mould on stems often indicates their presence.

Treatment is most effective in late spring and early
summer, when aphid populations are active. Pyrethrum-
based sprays or insecticidal plant oils and fatty acids (see
How to treat your plants with insecticidal soaps, invigorators
or oils, page 160) can help manage infestations, although
reapplication is necessary. Applying a balanced fertilizer
with nitrogen and potassium in early autumn can aid plant
recovery. Regular monitoring and early intervention are key
to minimizing damage.

Dutch elm disease *Ophiostoma novo-ulmi*

PLANTS AFFECTED
Elms (*Ulmus*)

DAMAGE CAUSED
Yellowing distorted foliage;
discoloured wood and bark;
premature tree death

Dutch elm disease (DED) is a deadly fungal disease that
is spread by elm bark beetles or infected wood. Symptoms
include wilting, yellowing and curling leaves, drooping twigs,
dark streaks beneath the bark, and dark spots in branch cross-
sections. The disease spreads through beetles carrying spores
and improper handling of infected tools, trees or logs. DED
originated in Asia and arrived in the UK from Canada in the
1960s. Control measures include: pruning diseased branches
early, using disinfected tools (see How to clean tools and other
equipment, page 104); removing infected elm wood; and
severing root grafts. Planting disease-resistant elm cultivars such
as *Ulmus* 'Wingham' can also help.

Typical Dutch elm disease damage.

How to clean tools and other equipment

Keeping your garden tools clean and well-maintained is essential for efficient gardening and preventing the spread of plant diseases. Pruning tools, for example, can transfer viral infections from one plant to another through contaminated sap. Likewise, soil-borne diseases can spread through infected dirt clinging to gardening tools.

Follow these steps to ensure your tools stay in top condition. By doing this, you'll extend the life of your tools and help maintain a healthy productive garden.

CLEANING CULTIVATION TOOLS

For tools such as spades, hoes, forks, rakes and trowels:

- Remove soil: Before leaving the site, clean off soil from the blades and shafts of cultivation tools, with a stiff scrubbing brush, to avoid transferring contaminants. For extremely muddy tools, rinse with a hose.
- Disinfect: Use a strong disinfectant or other sterilizing solution to clean all cultivation tools.
- Dry thoroughly: Dry each tool with an old towel. Drying prevents wooden handles from swelling and protects metal from rusting.
- Apply oil: Traditionally, tools were oiled with general-purpose oil to prevent rust. Even though many modern tools are made of stainless steel and are rust-resistant, occasional oiling may still be necessary, depending on metal quality.

CLEANING CUTTING TOOLS

For tools such as secateurs, knives, loppers, pruning saws and shears:

- Remove sap and grime: Dried sap can

attract dirt and reduce performance. Use a pan scourer or wire wool to scrub the blades.

- Lubricate: Apply penetrating oil or water-displacing lubricant spray to loosen any remaining grime. Wipe off residue with a clean cloth.
- Sterilize blades: Wipe blades with a strong disinfectant, then leave for 15–20 minutes to kill pathogens. Remove disinfectant with a towel.
- Tighten and oil: Check for loose parts, tighten if needed, and oil the central pivot point, with general-purpose oil. Open and close the tool to distribute the oil evenly.
- Store properly: Keep tools in a clean dry place to prevent rust and corrosion.

DON'T FORGET YOUR BOOTS AND CLEANING MATERIALS

Contaminated soil on boots, brushes and cloths can also spread diseases.

- Remove soil and disinfect the soles of your shoes or boots after working in potentially contaminated areas.
- Clean brushes and cloths thoroughly.

Left: Dirty tools and gloves can harbour diseases and fail to work optimally.
Above: It is essential that you keep all gardening equipment clean in order to reduce the risk of spreading infection.

Red band needle blight *Dothistroma septosporum*

PLANTS AFFECTED
Coniferous trees, especially pines (*Pinus*)

DAMAGE CAUSED
Premature needle drop; stunted growth

Red band needle blight is a significant disease that primarily affects conifer trees and is of major economic concern due to its impact on timber yields. It causes premature needle drop, leading to defoliation, reduced growth and, in severe cases, tree death. It is often confused with diplodia tip blight (see page 108). Pines are particularly vulnerable, with species in the UK such as Austrian pines (*P. nigra*), shore pines (*P. contorta*), Scots pines (*P. sylvestris*), western yellow pines (*P. ponderosa*) and bishop pines (*P. muricata*) being commonly affected. Recently, other conifer species, including larches (*Larix*), firs (*Abies*), hemlocks (*Tsuga*), spruces (*Picea*) and Douglas firs (*Pseudotsuga menziesii*), have shown low levels of susceptibility.

Trees of all ages can be attacked, with symptoms typically appearing first at the base of the crown on older needles. Infected needles develop yellow and tan spots that progress to red bands. As the disease advances, the needle ends turn reddish brown while the bases remain green. Within the red bands, small black fruiting bodies form, releasing spores in early and midsummer.

Above: Typical red band needle blight symptoms.

These spores infect the current year's needles, which later shed, leaving branches with a characteristic 'lion's tail' appearance – a tuft of new needles at the branch tips. This annual cycle of defoliation weakens trees, significantly reducing timber yields and potentially leading to tree death.

The fungus spreads under moist conditions, requiring water for natural dispersal. Spores are carried by wind, rain or mist, enabling long-distance spread. Movement of infected material, including needles on footwear, clothing, machinery or timber, can also extend the disease.

Unfortunately, there are currently no fungicides available in the UK to control red band needle blight. However, good management practices can help reduce its impact. These include clearing and burning fallen needles, to minimize spore sources, and removing heavily infected trees to prevent further spread. Maintaining good tree health, such as by ensuring optimal growing conditions and by reducing stress, may also help trees resist infection.

This disease poses a growing threat, particularly with climate change bringing warmer wetter conditions that favour its spread. Effective monitoring and prompt management are essential to minimize losses and preserve affected tree populations.

Diplodia tip blight *Diplodia sapinea* aka *Sphaeropsis sapinea*

PLANTS AFFECTED
Mainly, two- and three-needled pines (*Pinus*)

DAMAGE CAUSED
Stunted discoloured shoot tips; resin bleeding; dead shoots

Diplodia tip blight, formerly known as sphaeropsis blight, is caused by a fungal pathogen that produces small, black, ovoid fruiting bodies on the needles containing conidiaspores. These spores are dispersed through water splash, enabling the fungus to infect pine (*Pinus*) needles and shoots. The pathogen can exist in multiple forms, acting as an endophyte within living tissue without causing visible symptoms, or as a saprotroph colonizing dead needle tissue. In recent years, this disease has become increasingly prevalent across the UK, posing significant challenges to forestry and ornamental tree management.

Two- and three-needled pines are the primary hosts, but other conifers are also susceptible, including Lawson's cypresses (*Chamaecyparis lawsoniana*), grand firs (*Abies grandis*), Norway spruces (*Picea abies*) and Douglas firs (*Pseudotsuga menziesii*). In warmer climates, species such as western yellow pines (*Pinus ponderosa*), Monterey pines (*P. radiata*) and Austrian pines (*P. nigra*) are severely affected. Additional species, including Scots pines (*P. sylvestris*), dwarf mountain pines (*P. mugo*), Mexican weeping pines (*P. patula*), lodgepole pines (*P. contorta*) and Canadian red pines (*P. resinosa*), have also been reported as vulnerable.

Only the current season's growth is susceptible to infection. Diseased shoot tips fail to elongate fully, turn yellow or brown, and may curl. Resin bleeding often occurs on infected shoots, while new growth may appear below the affected area. Dark brown or purple lesions develop along the current year's stems, spreading up to the junction with the previous year's growth. Needles typically turn reddish brown and then grey as the disease progresses. Fruiting bodies form at the bases of infected needles, primarily in late spring or summer, although they can occasionally develop in autumn following infection. These fruiting bodies release spores in water droplets, which germinate on current-year needles under favourable conditions, with optimal temperatures around 24°C/75°F.

Symptoms of needle and shoot death appear in summer. The fungus overwinters in dead needles, bark, wood and cones, allowing the infection cycle to restart the following spring. When diplodia tip blight occurs over successive years, dead shoots and branches become visible throughout the crown, leading to significant reductions in growth and, in severe cases, tree mortality.

Effective management focuses on reducing stress, particularly drought stress, which exacerbates susceptibility. Thinning dense stands to improve airflow can reduce humidity levels and hinder spore dispersal and germination. Collecting and burning fallen needles, along with removing heavily infected trees, can further limit the spread of diplodia tip blight. No fungicides are currently approved for use in the UK.

Oak processionary moths *Thaumetopoea processionea*

PLANTS AFFECTED
Oak (*Quercus*)
DAMAGE CAUSED
Defoliation

The oak processionary moth (OPM) presents a significant threat as its caterpillars can cause extreme damage, both to oak (*Quercus*) trees and to the health of humans and animals in close proximity. These caterpillars possess tiny hairs containing an irritating protein called thaumetopoein, from which the species derives part of its scientific name. Contact with these hairs can result in itching skin, eye irritations, sore throats and breathing difficulties, with the risk of exposure being highest in late spring and early summer.

Caterpillars can shed these hairs when threatened or when disturbed, and wind can disperse them, leading to accumulation in caterpillar nests, which may fall to the ground or cling to any surrounding vegetation and clothing. It's crucial to avoid touching or approaching OPM nests or caterpillars. This is especially important for children, pets, individuals working near oak trees and anyone spending time close to infested trees, particularly on windy days in summer, as well as for grazing and browsing livestock and wild animals. If contact with OPM occurs, seek advice from a pharmacist for relief from skin or eye irritations, and if serious allergic reactions are suspected contact emergency health services or see a doctor, specifying OPM exposure. Should pets or other animals be affected, consult a veterinary surgeon.

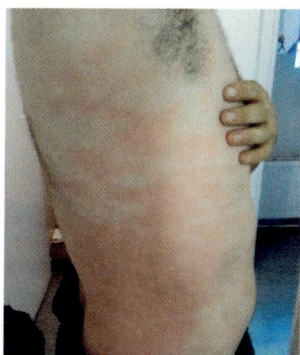

Top, left: OPM caterpillars.
Top, right: Oak processionary moth nest.
Above: Torso rash triggered by contact with oak processionary moths.

Commonly mistaken moths for OPM include the European gypsy moth (*Lymantria dispar*), brown-tail moth (*Euproctis chrysorrhoea*), buff-tip (*Phalera bucephala*), vapourer (*Orgyia antiqua*) and small eggar (*Eriogaster lanestris*). OPM caterpillars feed on oak foliage, and large populations can defoliate entire trees, rendering them more susceptible to other pests, diseases and stresses such as drought.

Suspected OPM nests or caterpillars should be reported immediately to official government authorities, such as TreeAlert in England, Scotland or Wales. Reports will be assessed by scientists and forwarded to plant health authorities for appropriate action, thereby aiding the management and containment of this significant threat to oak trees and public health.

Control OPM with an insecticidal spray and, in some cases, by removing caterpillar nests. The caterpillar should be targeted in the early stages of its life cycle, in late spring, by spraying. Once the caterpillars are pupating, then nest removal is required, in midsummer. Both measures should be executed by professionals equipped with appropriate training and gear.

Acute oak decline

PLANTS AFFECTED
Oaks (*Quercus*)
DAMAGE CAUSED
Bark; thinning tree canopy;
dead wood

Acute oak decline (AOD) is an emerging disease that was first observed in the UK in the late twentieth century. It can kill oak (*Quercus*) trees within 4–6 years of symptom onset, and it predominantly affects mature oaks, although younger trees are susceptible, too. The disease is caused by multiple agents, particularly bacteria, and its onset is often linked to environmental stressors that predispose trees to infection. Thousands of trees in the UK have been impacted.

The disease is most prevalent in warm, drought-prone regions of south-eastern, central and eastern England, as well as the Welsh Borders and south-east Wales. Environmental factors, including airborne nitrogen pollution and low dry sulphur levels, are believed to contribute to its spread. As of 2020, AOD had not been reported in Scotland or Northern Ireland. Globally, similar oak decline diseases have been observed in continental Europe, western Asia and the Americas, highlighting AOD as a worldwide concern that may have been overlooked because of its complex causative agents.

In the UK, AOD primarily affects the two native oak species: English oak (*Quercus robur*) and sessile oak (*Q. petraea*). However, other oak species, including holm oak (*Q. ilex*), Pyrenean oak (*Q. pyrenaica*) and red oak (*Q. rubra*), have also been affected. Mature trees over fifty years old are most impacted, but younger trees with stem diameters as little as 10–12cm/4–5in have also shown symptoms.

Symptoms of AOD include dark-coloured, vertical, weeping fissures on

Right: Acute oak decline fissure symptom.

the trunk, known as cankers. These bleed black fluid through vertical cracks in the bark and can dry into caked deposits. Beneath these bleeds, lesions form in the live tissue. However, cankers alone do not confirm AOD, as they can also result from attacks by other pests or pathogens, such as *Phytophthora* species (see Phytophthora root rot, page 98, and Blight, page 24). In about one-third of cases, D-shaped exit holes made by two-spotted oak buprestid beetles (*Agrilus biguttatus*) are visible in the bark. Larval galleries from these beetles are often found beneath the lesions.

As the disease progresses, the tree canopy thins significantly, often within eighteen months, and rapid decline may follow. AOD poses a severe threat to oak trees, which play vital roles in the UK's economy, environment and biodiversity, supporting more species than any other tree. No treatment has been identified as yet.

Oak pinhole borers *Platypus cylindrus*

PLANTS AFFECTED
Oaks (*Quercus*) and some
other hardwoods
DAMAGE CAUSED
Extensive tunnels in the
timber

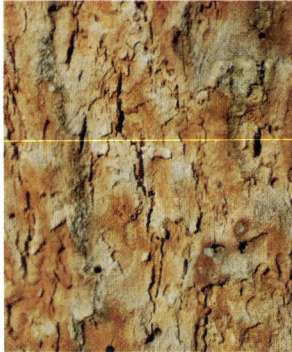

Top: Adult oak pinhole borer.
Above: Typical oak pinhole
borer damage.

Although the oak pinhole borer is a pest primarily affecting oak
trees (*Quercus*), it can also infest other hardwoods, including
sweet chestnuts (*Castanea*), beeches (*Fagus*), ash (*Fraxinus*),
elms (*Ulmus*) and walnuts (*Juglans*). Its larvae bore into the
heartwood of stressed or dying trees, damaging the timber and
reducing its value. It is the only borer capable of tunnelling into
oak heartwood without fungal decay.

The adult beetle is 6–8mm/¼–⅜in long, pitch-brown
to black, with a rectangular body and prominent head. It is
larger than an oak bark beetle (subfamily Scolytinae) and has
an elongated pronotum. The beetle is native to the UK and
continental Europe, targeting weakened or dead trees but not
killing healthy ones. Its larvae create galleries approximately
1.6mm/¹⁄₁₆in in diameter, round and free from bore dust. These
tunnels often stain the surrounding wood black or brown,
because of ambrosia fungi. Initially, the tunnels run across
the grain, branching in various directions as the infestation
progresses. Although it does not kill trees, the oak pinhole
borer significantly affects timber quality, emphasizing the
importance of minimizing tree stress to prevent infestations.
There is currently no treatment available.

Horse chestnut leaf miners *Cameraria ohridella*

PLANTS AFFECTED
Horse chestnuts (*Aesculus*)
DAMAGE CAUSED
Leaves

Top: Typical leaf damage from horse chestnut leaf miners.
Above: Horse chestnut leaf miner larva.

The horse chestnut leaf miner is the larva of the moth *Cameraria ohridella*. Adult moths are diminutive, measuring 4–6mm/ ⅙– ¼in in length. During early summer, the adult female deposits up to 180 eggs on newly unfurled leaves on horse chestnut trees (*Aesculus hippocastananum*). The hatched larvae commence feeding on the leaves, progressing through several growth stages. As they feed, they create elongated white patches on the leaves, which later turn brown. This is caused by the larvae 'mining' through the leaves. In summer, inspecting affected leaves held up to sunlight might reveal the tiny caterpillars or their circular pupal cocoons within the mined areas. Following their growth stages, the larvae undergo pupation.

While foliage spray can manage this pest, one of the most effective controls is to collect the leaves in autumn and incinerate them. This act reduces the population of leaf miners overwintering in the leaves, thereby mitigating their impact on horse chestnut trees.

How to use low-tech methods in practical solutions for soil compaction

Soil compaction initiates a cascade of problems for trees, hindering root growth and impeding essential air and water movement. This restriction leads to physiological dysfunctions, impairing systemic functions and the tree's ability to adapt to environmental changes. Compacted soil disrupts respiration processes, reducing water absorption and causing leaf water deficit. Hormonal imbalance, including increased levels of abscisic acid and ethylene, further exacerbates stress.

Moreover, compaction diminishes absorption of major mineral nutrients and reduces leaf area, thereby impairing photosynthesis productivity. Continuous compaction drives roots into an anaerobic state, disrupting respiration further.

These low-tech solutions offer effective means to address soil compaction.

AERATE AND MULCH

Aeration, employing a garden fork to break up any compacted soil surface, facilitates pore space opening by penetrating the soil and moving it back and forth. This technique, when combined with mulching, not only adds nutrients but also increases soil activity, aiding in de-compaction.

Mulching alone can gradually improve soil compaction, making it ideal for over-mature or veteran trees as it does not disturb roots. Fencing off trees, to prevent traffic within the root zone, is particularly beneficial when leaves are left on the ground in autumn, and it creates conditions akin to the woodland floor, thereby promoting tree health.

Aeration techniques, such as perforating soil with spikes or coring devices, enhance air and water penetration, assisting soil de-compaction.

RETHINK YOUR PLOT

In garden settings, altering the mowing regime can be highly beneficial. Leaving areas of long grass within the rooting environment of a tree and planting wild flowers or bulbs can create barriers that reduce traffic there. Allowing brambles and other woody plants to establish under trees can also help mitigate compaction (see How to establish young trees and shrubs, page 126).

AVOID HEAVY MACHINERY

In large areas of parkland or estates, practices like rotational grazing minimize soil disturbance by livestock, while avoiding heavy machinery on wet soil prevents compaction. Promoting biodiversity through planting cover crops and perennial vegetation further enhance soil health and structure.

Regular monitoring and assessment of soil compaction levels are essential for timely interventions.

By employing these low-tech methods and adopting sustainable land management practices, you can mitigate the detrimental effects of soil compaction on trees and promote healthy ecosystems. Nurturing soil health ensures the vitality and longevity of trees while fostering biodiversity and ecosystem resilience. Thus, implementing these practices contributes to the overall well-being of trees and the environment.

Opposite: Use a garden fork to gently lift and loosen the soil, so air and water can reach plant roots.
Above: Changing your mowing regime can greatly benefit soil health by reducing compaction and allowing vegetation to recover.

Oriental chestnut gall wasps   *Dryocosmus kuriphilus*

PLANTS AFFECTED
Sweet chestnuts (*Castanea*)
DAMAGE CAUSED
Buds, leaves and petioles;
premature leaf drop

Above: Typical gall
growth from oriental
chestnut gall wasps.

The oriental chestnut gall wasp, a small species of wasp, lays larvae that induce abnormal growths known as galls on sweet chestnut trees (*Castanea*), affecting buds, leaves and petioles (leaf stalks). High gall infestations weaken trees, making them more vulnerable to sweet chestnut blight (see opposite) and potentially leading to tree decline.

Initially green, rose-pink or red, fresh galls dry out and become woody over summer, coinciding with adult wasp emergence. This premature drying triggers early leaf drop, with galls typically shedding in autumn, although those attached to petioles may persist for up to two years.

Control methods mainly rely on the parasitic wasp *Torymus sinensis*, which preys on the gall wasp. Despite its presence in England, *T. sinensis* numbers are low. In April 2021, Fera Science Ltd released *T. sinensis* at various sites to boost its population and effectively combat the gall wasp pest.

Sweet chestnut blight *Cryphonectria parasitica*

PLANTS AFFECTED
Sweet chestnuts (*Castanea*)
DAMAGE CAUSED
Bark; brown wilted leaves

Typical sweet chestnut blight damage.

This is a serious fungal disease of sweet chestnut trees (*Castanea*) and was first reported in the UK in 2011, primarily in central and southern England. The fungus infects bark through wounds, forming sunken cankers and orange stromata that release spores in moist weather. Above the canker, leaves wilt and turn brown, while healthy foliage and new shoots may appear below it. Infected plants should be destroyed on site. Biosecurity measures, such as disinfecting tools and equipment (see How to clean tools and other equipment, page 104), are essential to prevent the spread of sweet chestnut blight.

Cherry blackfly *Myzus cerasi*

PLANTS AFFECTED
Cherry (*Prunus*)
DAMAGE CAUSED
Tightly curled, crumpled leaves; sticky honeydew; premature leaf drop

Typical cherry blackfly damage.

This aphid feeds on cherry leaves, causing them to curl inwards for protection. Honeydew deposits can lead to sooty mould, while affected leaves may turn yellow and drop. Although cherry blackfly is best tolerated without insecticides, severe infestations may require intervention.

Treatment involves pruning heavily affected growth in midsummer and applying a winter tree wash to eliminate overwintering aphids. Biological controls, such as introducing beneficial predators, can also help manage populations naturally (see How to use biological controls, page 90). To prevent infestations, regularly monitor for pests and apply organic oil-based miticides or insecticides every 7–10 days to protect the tree.

Ash dieback *Hymenoscyphus fraxineus* aka *Chalara fraxinea*

PLANTS AFFECTED
European ashes (*Fraxinus excelsior*)

DAMAGE CAUSED
Blackened wilted foliage; tree death

This devastating fungal disease poses significant threats to the UK's ash (*Fraxinus*) population, biodiversity and hardwood industries. Native to eastern Asia, ash dieback has co-evolved with Asian ash species like Chinese ash (*F. chinensis*) and Manchurian ash (*F. mandshurica*), which show tolerance to the infection. In contrast, in Europe the disease primarily affects European ash (*F. excelsior*), including the ornamental weeping ash (*F.e.* 'Pendula'), as well as narrow-leaved ash (*F. angustifolia*). Manna ash (*F. ornus*) appears less affected, but all ash species worldwide are considered susceptible to some extent.

Effective identification of ash dieback requires careful observation. Symptoms include leaves wilting and blackening between midsummer and early autumn, and premature leaf shedding. As the infection spreads, dark, diamond-shaped lesions form on bark, often around branch joints, spreading along branches and trunks; stems dry out and crack over time, eventually girdling the tree and disrupting nutrient flow.

Above, left: Diseased trunk caused by ash dieback.
Above, right: Typical leaf damage from ash dieback.

Young and coppiced ash trees are particularly vulnerable, often dying quickly, while mature trees may resist for years until weakened by prolonged exposure or secondary infections such as honey fungus (see page 100).

Misidentifying the disease is common, as ash trees exhibit natural seasonal and genetic variations. In spring, ash trees are among the last to come into leaf, and variations in leafing times between individual trees or regions may be mistaken for disease. By midsummer, healthy ash trees should have fully developed leaves. In autumn, retained clumps of dark-coloured keys (seeds) can resemble diseased foliage from a distance. Other causes, such as environmental stress or pests, can also lead to shoot death, complicating diagnosis. Although some infected trees may recover temporarily, most succumb eventually.

Management involves removing and destroying infected trees, planting disease-resistant cultivars and limiting fungal spread through biosecurity measures (see How to clean tools and other equipment, page 104).

Sirococcus blight *Sirococcus tsugae*

PLANTS AFFECTED
Various cedars (*Cedrus*)
and hemlocks (*Tsuga*)
DAMAGE CAUSED
Severe shoot blight
and defoliation

Initially identified in western North America, this fungal disease has since been reported in south-eastern (Georgia) USA and the north-eastern USA. In the UK, it was first detected in 2014, with confirmed cases in England, Scotland, Wales and later Northern Ireland. The spread of this fungus has raised concerns about its potential to cause substantial damage to ornamental trees in gardens and parks, as well as economic losses in the nursery sector.

Several species are susceptible to the disease, including Atlas cedars (*Cedrus atlantica*), Himalayan cedars (*C. deodara*), cedars of Lebanon (*C. libani*), Cyprus cedars (*C. libani* var. *brevifolia*), western hemlocks (*Tsuga heterophylla*), mountain hemlocks (*T. mertensiana*) and eastern hemlocks (*T. canadensis*). Interestingly, eastern hemlocks seem to exhibit greater tolerance when compared to western hemlocks.

The symptoms of sirococcus blight are most visible in spring. Affected cedar trees display dead needles, which are distinctive due to their initial pinkish coloration before turning brown as the season progresses. Additional symptoms include dead shoots, cankers and gum exudation. Stromata of sirococcus blight can often be found on dead needles. Infected branches may exhibit subtle cankers, marked by reduced branch diameter and a colour change in the bark from green to dark red or purple. Resin bleeding is another common sign, and brown lesions in the bark's phloem tissue can spread from the shoots to branches and even the trunk, where they extend longitudinally. On western hemlocks, the disease is particularly evident in natural stands within the understorey, often affecting multiple shoot tips on a single tree. Mountain hemlocks commonly show symptoms of shoot blight caused by the fungus.

Sirococcus blight spreads locally through rain splash, with strong winds potentially dispersing spores over longer distances. While seed transmission has been documented for related species, no evidence currently supports seed transmission of this fungus. Planting stock, cut foliage and seeds from infected regions represent potential pathways for the fungus's spread, posing risks to international and domestic movement of cedar and hemlock trees. The spread of sirococcus blight underscores the need for vigilance, early detection and stringent biosecurity practices to safeguard cedar and hemlock species.

Unfortunately, no effective control measures for sirococcus blight have been reported in forests, and limited information exists on managing the disease in nurseries or gardens. Current strategies focus on biosecurity and plant hygiene (see How to clean tools and other equipment, page 104) to reduce the risk of spread. Preventative measures include sourcing clean planting stock, avoiding imports from infected areas, and maintaining proper sanitation and spacing in managed landscapes.

Bacterial canker *Pseudomonas syringae* pv. *morsprunorum* and *P. syringae* pv. *syringae*

PLANTS AFFECTED
Plums, cherries and
other *Prunus* species
DAMAGE CAUSED
Bark, stems and leaves

Bacterial canker is a disease caused by two closely related bacteria and typically begins in mid-spring, when cankers form on the bark, often followed by dieback of shoots. By early summer, affected foliage develops 'shotholes', with small, rounded, brown spots that eventually fall out, creating holes resembling shotgun pellets. This symptom gives the disease its popular common name, shothole. Sunken dead areas of bark also develop in spring and early summer, usually accompanied by a gummy ooze. However, it's important to note that gummosis (gum production) from the bark of *Prunus* species is common and can be caused by factors other than bacterial canker: for example, by physical damage or environmental stresses.

If the infection spreads around the branch, it can cause rapid dieback. Infected shoots may either fail to emerge or initially grow normally before rapidly dying back. In severe cases, this dieback can affect a significant percentage of shoots on a tree. Shoot dieback can also result from fungal diseases like blossom wilt, which can look like bacterial canker (see page 124).

Not all trees affected by bacterial canker will develop damaging cankers on the trunk or main branches. Many trees will exhibit leaf and shoot symptoms without showing more severe canker formation. The extent of these symptoms can vary

Above, left: Shotholes on leaves infected with bacterial canker. *Above, right*: Typical bacterial canker bark damage.

each year, influenced by weather conditions. Wet years tend to favour the disease, often leading to a sparse crown. In contrast, during drier growing seasons, affected trees may show much less damage.

To manage bacterial canker, it's essential to prune trees properly. Ideally, pruning should be done in mid- or late summer, when tree tissues are most resistant to infection. This timing also reduces the risk of infection by spores of the fungus responsible for silver leaf disease (see page 54). Prune out all cankered areas, removing them back to healthy wood. It's important to dispose of pruned material by burning or landfilling, to prevent the spread of the bacteria.

Some cherry and plum cultivars, such as 'Merton Glory', 'Merton Premier', 'Merla' and 'Warwickshire Drooper', have shown some resistance to bacterial canker. Selecting such resistant cultivars and practising good tree care can help minimize the impact of this disease.

How to establish young trees and shrubs

Planting trees and shrubs in your garden is a rewarding way to enhance the landscape, improve biodiversity and contribute to the environment. Follow this guide for successful planting and aftercare.

PLANNING YOUR PLANTING

- Site considerations: Assess the site for the tree's long-term growth, considering its final size in 10–100 years, as appropriate. Avoid planting too close to buildings, infrastructure or under other trees.
- Soil type: Test the soil for its composition – sandy, loamy or clay. Unless it is free-draining, deal with waterlogging concerns.
- Underground utilities: Check for utilities like gas and water lines by digging trial holes or scanning the site.
- Permissions: Obtain landowner approval before planting, if relevant.

SELECTING THE RIGHT TREE

- Purpose: Choose a tree species based on the desired outcome – is it for shade, ornamental value or biodiversity enhancement?
- Space and conditions: Match the tree to the available space, soil conditions and local climate.
- Diversity: Incorporate native and non-native species to mitigate climate change and disease risks.
- Size and rootstock: Decide on the size and type of tree (see box, opposite).
- Biosecurity: Source trees from reputable nurseries with strict biosecurity practices to minimize pests and diseases.

PLANTING THE TREE

Plant between mid-autumn and early spring for the best results, avoiding times when deciduous trees are in leaf.

- Prepare the hole: Dig a hole twice as wide as the root ball but no deeper. Save the topsoil to backfill.
- Check the depth of the hole: The soil mark on the tree stem should match the final ground level.
- Formative pruning: Remove damaged or rubbing branches before planting. Keep the main leader intact.
- Set and backfill: Place the tree in the hole, ensuring it is straight. Backfill with excavated soil and compact gently.

Above: Lie a plank across the planting hole to check its depth.
Right: Mulch round the newly planted tree.

TYPES OF TREE ROOT SYSTEM	OPTIMUM PLANTING TIME	COMMENT
Bare-root	Mid-autumn to early spring	Affordable but limited in species
Root-balled	Autumn to spring	Retain fibrous roots but require careful handling
Containerized	Year-round	Easy to plant but more expensive

SECURING THE TREE
- Protecting: Surround a tree with a girth of less than 10cm/4in with four timber posts and netting. Use two posts and rubber ties for a tree with a larger girth.
- Tying: Secure the netting or tree to each post. Position ties at one-third up the tree's height and ensure the tree can move slightly to strengthen its stem.
- Removing stakes and ties: Do this once the tree has established.

AFTERCARE
- Mulching: Apply 5–10cm/2–4in of bark mulch in a 1m/3ft diameter around the tree. Keep mulch away from the stem to prevent rot.
- Watering: Water thoroughly after planting and during dry spells.
- Community Engagement: Involve the local community and leave nursery tags in place for tree identification.

Anthracnose (of plane) *Apiognomonia veneta*

PLANTS AFFECTED
Planes (*Platanus*)
DAMAGE CAUSED
Leaves, twigs and buds

Above: Typical anthracnose leaf damage.

One of the most susceptible species within the *Platanus* genus affected by anthracnose is western plane (*P. occidentalis*), also known as American sycamore, while oriental plane (*P. orientalis*) shows greater tolerance. Their hybrid, London plane (*P. × hispania*), has variable susceptibility depending on the clone. Anthracnose is widespread in the UK, continental Europe, the USA, Russia and New Zealand, thriving particularly in mild wet springs. The disease symptoms, appearing in spring, include bud blight, twig dieback and cankers on branches, followed by shoot and leaf blight. Affected leaves display brown stains along veins and crinkled patches, which can be mistaken for frost damage but differ because of the accompanying twig cankers and wood staining. Although heavily infected trees can defoliate and appear unsightly, they typically recover by midsummer, with long-term effects generally minor. However, repeated infections can reduce vigour and distort branch growth.

Management focuses on cultural practices, such as removing infected leaves and pruning affected branches, as chemical treatments are rarely used in the UK. Thinning tree crowns to improve airflow can also reduce disease severity. Biosecurity measures, including cleaning tools and other equipment (see page 104), are vital to prevent the spread, while selecting anthracnose-tolerant London plane cultivars may aid disease management.

Coryneum canker *Seiridium cardinale*

PLANTS AFFECTED

Leyland cypresses
(× *Hesperotropsis leylandii*,
aka × *Cuprocyparis leylandii*),
Monterey cypresses
(*Hesperocyparis macrocarpa*,
aka *Cupressus macrocarpa*)
and western red cedars
(*Thuja plicata*)

DAMAGE CAUSED

Stems and leaves; canopy
thinning; weakened growth

Above: Typical coryneum
canker damage.

Coryneum canker, also known as cypress canker, is a serious fungal disease. It causes browning or reddening stems, sunken, dark brown cankers, black pustules on twigs, grey foliage and gummy oozing lesions. As the disease progresses, branches die back, leading to canopy thinning and overall decline. The fungus spreads through rain, wind, birds, insects and contaminated pruning tools, making it difficult to control. Wet conditions, poor airflow and environmental stress increase susceptibility. Once established, coryneum canker can be persistent, requiring careful management to reduce its impact.

Early detection, good hygiene and proactive care are the best approaches for managing this disease in the UK, because suitable fungicides to control it are not permitted. Increased plant spacing, to enhance airflow and avoid excessive moisture accumulation, is key to prevention, too (see How to establish young trees and shrubs, page 126).

To help reduce disease severity, prune infected branches in dry weather, using properly disinfected tools (see How to clean tools and other equipment, page 104), and improve tree vigour through mulching, watering and fertilization. Encouraging healthy growth by tying in new shoots to fill gaps and selecting resistant cultivars can also aid long-term recovery.

HOUSEPLANTS

BY PAUL REES

Keeping your houseplants happy starts with good cultivation. Plants growing in their appropriate conditions, with enough food, water and light, will be less susceptible to pests and diseases. To ensure they are healthy, monitor plants regularly, especially ones recently added to your collection. Any outbreak is much easier to manage if it is spotted early.

Ants *Lasius niger* (Formicinae), *Monomorium minimum* (Myrmicinae) and *Technomyrmex albipes* (Dolichoderinae)

PLANTS AFFECTED
Wide range of plants attracting honeydew pests or producing sticky plant residues

DAMAGE CAUSED
Ants support pests such as scale, mealybugs and aphids

Ants are found throughout the world, and some species are among the most invasive animals we know. They are not generally considered plant pests, despite nesting in pots, because they cause little damage to the plant (unless you have a colony of leafcutter ants, that is). There are several ants that farm pests, such as the European black garden ant (*Lasius niger*), which may enter a greenhouse but will nest outdoors, whereas species originating outside of Europe like the little black ant (*Monomorium minimum*) and the white-footed ant (*Technomyrmex albipes*) are more at home in a greenhouse.

The many ant species within the subfamilies Myrmicinae, Dolichoderinae and Formicinae are known to harvest honeydews (sugar-rich secretions deposited on leaves and stems) and have a mutual relationship with honeydew-producing insects like scale insects (see page 144), mealybugs (see page 140) and aphids (see page 134). The ants in effect farm these insects. They move them around when they are overcrowded and protect them from predators, all in return for honeydew, a sweet surgery reward. To control scale insects, mealybugs and aphids you need first to combat ants; otherwise, any beneficial insects released will be attacked by the ants, or the ants will quickly rebuild the farm after spraying.

The best way to combat ants is to beat them at their own game. They love sweet substances. Putting a drop of honey where there are ants will quickly attract the scout ants. When these scouts exit from the nest in search of new food sources, they each leave a pheromone trail on their way. If they find food, they follow the trail back, thereby

strengthening the pheromone on their return. This allows the foraging ant to follow the trail to the food source. The more ants that follow it, the stronger the trail gets. When the food has run out, they stop reinforcing the trail when returning to the nest, letting the scent get weaker and weaker.

Using a sweet substance like honey will hopefully get the colony trailing from the honey back to the nest, at which point it is easier to follow them back and find where they are nesting. Once you know where the nest is, place bait stations close by. These bait stations have sugar mixed with a pesticide. The ants will collect it and take it back to the nest, feed the queen and cause the colony to collapse. Placing the bait stations as close as possible to the nest is important, as if the bait starts to make the ant ill, or kill it before it enters the nest, it will warn the others not to feed from the bait station.

If you find ants are nesting in a pot plant, the easiest way to get rid of them is to repot the plant and discard the nest with the old potting compost. There are also ant dusts and powders you can use. However, these tend to be less effective in damp conditions and target only the foraging ants and not the whole colony.

Aphids (blackfly and greenfly) *Aphis fabae* and *Aulacorthum solani*

PLANTS AFFECTED
Wide range of ornamental
plants and crops
DAMAGE CAUSED
Distorted leaves and
weakened plants

Aphids start as small wingless nymphs, most often on the
fresh tips of your plants. They will moult three or four times
until they eventually form adults. They are sap-sucking,
feeding on sugary sap in the phloem, thereby depriving the
plant of nutrients needed to grow. Aphids will typically start
feeding at the growing tips of the plants, causing the leaves to
curl and become distorted. Large infestations can severely
weaken and even kill the plant in extreme cases.

When feeding, an aphid inserts its stylet into the plant
and 'regurgitates' some of the sap from a previous
plant it was feeding on. This feeding habit makes
aphids the perfect vector for spreading plant
viruses from one plant to another. They also
produce copious amounts of honeydew,
which serves as a reward for ants (see page
132); however, the honeydew that is not
collected by ants leaves a sticky residue on
plants. These are perfect places for sooty
mould (see page 156) to grow.

Aphids travel on the wind and will enter a
greenhouse through the vents. Opening the
leeward vents and keeping the windward
vents closed will cause aphids to be
blown over the greenhouse rather
than in it. Aphids indoors can
enter through open windows
or on clothing.

Controlling aphids can be
tricky as small populations can
rapidly increase through their
ability to reproduce asexually.

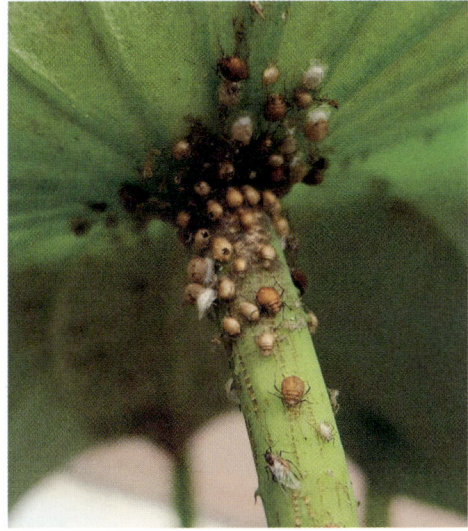

Above, left: A colony of blackfly with various instars (stages).
Above, right: Parasitized aphids.

This quick rise in numbers makes regular inspections and early detection invaluable. Use sticky traps to help monitor for new arrivals (see How to catch pests with sticky friends, page 138). A small population can be easily pruned out and discarded. If pruning is not possible, spray plants with a mixture of 5ml/ 1 tsp dishwashing liquid in a 500ml/17½fl. oz bottle or with a plant invigorator to reduce the population; do this regularly.

Many beneficial insects can help control aphids. Hoverflies (Syrphidae family) lay their eggs in among the colony, and their larvae will feed on aphids. Similarly, the aphid midge (*Aphidoletes aphidimyza*), another predatory fly with a smaller orange larva, can be used. Biologicals such as lacewings (*Chrysoperla carnea*) or ladybirds (*Adalia bipunctata*) can be ferocious predators. Parasitic wasps such as *Aphidius colemani* and *A. ervi* species can be extremely useful, too. Unless you know which species of aphid you have, it is best to use a mix of different parasitic wasps as some prefer some aphid species over others. When using these wasps, allow them to establish before resorting to other means of treatment. See also How to use biological controls, page 90.

Broad mites *Polyphagotarsonemus latus*

PLANTS AFFECTED
Wide range of ornamentals including *Begonia, Chrysanthemum*, dumb cane (*Dieffenbachia*), *Fuchsia, Gerbera, Gloxinia, Hibiscus,* busy Lizzies (*Impatiens*), *Pelargonium, Peperomia* and African violets (*Saintpaulia*)

DAMAGE CAUSED
Distorted, crinkled and stunted new growth with downward-curling leaves; severe cases lead to discoloration, interveinal chlorosis and dying leaves

Above: Typical broad mite damage.

The broad mite is a microscopic, sap-sucking pest that feeds on the undersides of new leaves, at their bases; this causes distorted new growth. (Mature leaves are unaffected.) These symptoms are similar to some viruses or physiological causes and are often misdiagnosed because broad mites can't be seen with the naked eye. The damage they do is believed to be caused by toxins in the mites' saliva, when they are most active in summer. The broad mite life cycle has four stages: egg, larva, nymph and adult. Clear eggs are laid on the underside of a leaf, and they hatch in 2–3 days. Adult male mites carry females in their final moult to new leaves. Further dispersion through the greenhouse is through clothing, tools and attaching to other vectors such as glasshouse whitefly (see page 166).

To control broad mites, prune out damaged growth and spray the plants in the vicinity with horticultural oils or soaps (see How to treat your plants with insecticidal soaps, invigorators or oils, page 160). Predatory mites such as *Amblyseius andersoni* and *Neoseiulus californicus* can be released to help control broad mite populations (see How to use biological controls, page 90).

Glasshouse leafhoppers *Hauptidia maroccana*

PLANTS AFFECTED
Wide range of greenhouse ornamentals and crops

DAMAGE CAUSED
Mottling with yellow or bleached white spots on upper leaf surfaces

Top: Glasshouse leafhopper nymph.
Above: Adult glasshouse leafhopper.

Leafhoppers are sap-sucking insects, 3mm/⅛in long, coloured light yellow or green with darker markings on the wings and distinct black spots on the head and thorax. Nymphs are wingless and are similar in appearance to aphids (see page 134). They feed on the undersides of leaves by inserting their stylets, located in their proboscis, and sucking out the contents, leaving empty, air-filled cells, which turn white or yellow, causing mottled symptoms. Females lay eggs in the leaf veins, which hatch ten days later. The nymphs go through five moults with the outer skin often remaining attached to the leaf, from where this pest gets its other name of ghost fly. In most cases, damage is not detrimental to the plant, although leafhoppers can transmit plant viruses.

Red sticky traps placed among foliage help monitor numbers (see How to catch pests with sticky friends, page 138). Releasing lacewing (*Chrysoperla carnea*) larvae will reduce infestations, too. Another way is to take the plants outdoors and spray them with a strong jet of water; this will remove most of the adult leafhoppers. Then treat the plants with horticultural soaps, invigorators or oils (see page 160), to control any remaining or newly emerging nymphs.

How to catch pests with sticky friends

Sticky traps are a great way to monitor flying pests. They can be hung among your plants or be mounted on to a pea stick. Yellow-coloured traps will attract several pests including aphids (see page 134), leaf miners (see Chrysanthemum leaf miners, page 70), fungus gnats (see page 143), thrips (see page 152) and glasshouse whitefly (see page 166), while blue-coloured traps are good for monitoring western flower thrips. Use red-coloured ones for glasshouse leafhoppers (see page 137). Sticky traps will catch some of the beneficial insects, too, so don't place them too close to where you release any biocontrols.

Pheromone lures are available for some species of thrip, citrus mealybug (see Mealybugs, page 140) and a variety of moth species. To maximize effectiveness, position

the lure in the centre of a flat trap and hang or mount it among your plants. For moths use a delta trap, also with the pheromone lure in the centre. Such traps not only let you know whether these night-flying insects are in the greenhouse, but they will also reduce the male population, resulting in fewer eggs being laid. However, it has been suggested that using these traps within the greenhouse could attract more males from outdoors, so some growers recommend hanging them outdoors, to encourage males to leave the greenhouse and/or to close vents at night when using them indoors.

More attractive alternatives to sticky traps and lures are carnivorous plants, such as Mexican long-spurred butterwort (*Pinguicula moranensis*) and cape sundew (*Drosera capensis*), which are very good at catching unsuspecting flying insects and are not too difficult to look after.

Opposite, above and *below*: Sit back and watch your traps catch some of those annoying pests flying around your plants. Check them regularly to see what they are trapping and how many. This will give you a good idea of what pests you have and whether the population is increasing or decreasing.

Below: A butterwort (*Pinguicula*) here growing among other plants has caught a number of fungus gnats and whiteflies with its sticky leaves.

Mealybugs Pseudococcidae family

PLANTS AFFECTED
Cacti, succulents and tropical/subtropical houseplants

DAMAGE CAUSED
Distorted leaves and weakened plants

These sap-sucking, soft-bodied insects have white filaments and waxy powder coating, from where they get their name 'mealy'. There are many species of mealybugs; however, the most common ones found on ornamental plants are citrus mealybugs (*Planococcus citri*), long-tail mealybugs (*P. longispinus*), obscure mealybugs (*P. viburni*) and root mealybugs (*Rhizoecus* spp.). Citrus mealybugs, obscure mealybugs and root mealybugs will form woolly egg clusters containing hundreds of eggs, whereas long-tailed mealybugs give birth to live young. The young crawlers easily move between plants, while the adults tend to stay in one place. Many mealybugs can reproduce asexually, quickly building up in numbers. Adult males are short-lived, winged individuals and rarely seen – and even absent in some species; they don't feed. Males can be attracted by pheromones (see How to catch pests with sticky friends, page 138).

Spotting mealybugs early is crucial in controlling them. For this reason, if you add any new plants to your collection keep them separate for a month or so and monitor them regularly.

Small populations can be removed from the plant using a paintbrush and methylated spirits; then spray the plant with water. Alternatively, wash them off with a strong jet of water. Mealybugs in large numbers cause leaf drop, so clear this mealybug-infested leaf litter as soon as possible.

There are a few biological controls available. The larval stage of the lacewing *Chrysoperla carnea* will feed on juvenile mealybugs, while the larvae and adults of *Cryptolaemus montrouzieri* – a type of ladybird – will also consume many species of mealybugs. The younger stages of *C. montrouzieri* are remarkably like mealybugs so be sure not to remove them, thinking they are the 'bad guys'. Various parasitic wasps can help control populations, too (see How to use biological controls, page 90).

Mealybugs also produce honeydew, which attracts ants. Any ants farming these bugs will fight off beneficial insects and move the mealybugs from one plant to the next. Tackling the ants (see page 132) will help control mealybugs.

If these methods are not working, try using a plant invigorator or horticultural soaps or oils (see page 160).

Top: A cluster of mealybugs.
Above: Long-tailed mealybug adults with young crawlers.

Millipedes and woodlice

Includes *Oxidus, Blaniulus, Cylindroiulus* millipede and *Armadillidium* woodlice species

PLANTS AFFECTED
Epiphytes

DAMAGE CAUSED
Roots; bulbs; de-structured potting media

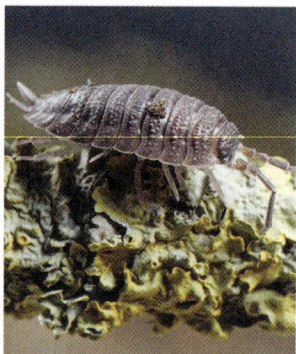

Top: Adult millipede.
Above: Adult woodlouse.

Millipedes and woodlice are generally considered as detritivores, feeding on dead plant material. They are, therefore, not typically considered a plant pest and are, in fact, beneficial in many ways as they help break down organic matter, thereby enriching soil. They can, however, feed on fine roots of seedlings and damage roots and tubers, especially if these roots are already damaged.

In houseplants, millipedes and woodlice are most problematic when using bark-based potting media, such as orchid potting composts. In large numbers, millipedes and woodlice will break down the bark component of the media, thereby reducing aeration and drainage, resulting in the plants needing more frequent repotting. Any orchids mounted on to cork bark will need replacing more frequently, too. Inspect new plants for millipedes and woodlice, and repot if necessary. Remove any fallen leaf litter as this can provide perfect conditions for these pests to thrive. Keeping your glasshouse clean and tidy goes a long way to reducing them, as does ensuring that unused potting mix is well sealed. If millipedes and woodlice are still problematic, drench pots with beneficial nematodes like *Steinernema carpocapsae* to help suppress them (see How to use biological controls, page 90).

Fungus gnat Sciaridae family

PLANTS AFFECTED
Seedlings and cuttings;
also mature cacti and
succulent plants

DAMAGE CAUSED
Root damage and rots;
weakened plants

Above: Adult fungus gnat.

Fungus gnats, also known as sciarid flies, are small winged flies, around 3mm/⅛in long, often seen flying near the surface of moist organic composts. They are attracted to the fungus in the potting compost, which breaks down such organic matter. The adults lay their eggs in the potting compost. The juveniles are small clear larvae, to 6mm/¼in long, with black heads. They feed initially on the fungus but later the fine roots of the plant. More problematically, fungus gnats can be vectors for root pathogens – a particular problem in cacti and succulents.

To avoid attracting fungus gnats, always buy good-quality potting soil and seal open bags. Also, do not overwater plants, as these flies are attracted to damp composts. Remove any rotting plants and repot badly infected compost. To help catch adults, set up sticky traps (see How to trap pests with sticky friends, page 138).

Useful biological controls are: predatory mites (*Hypoaspis*), which live in the soil feeding on the larvae; nematodes such as *Steinernema*; and predatory rove beetles such as *Atheta coriaria* (see How to use biological controls, page 90). It is worth remembering that *Atheta* and *Hypoaspis* will feed on each other so release only one at a time. Entomopathogenic fungi such as *Beauveria* species also provide some control.

Scale insects *Saissetia coffeae, Coccus* and *Diaspis* species

These small, soft-bodied, white, brown or black insects often feed near the veins of leaves. The females start out as 'crawlers' moving around to find a suitable place to feed; once feeding, they become immobile and form a protective, scale-like covering over their bodies. The male crawlers become winged and die soon after mating. Females can lay hundreds or even thousands of eggs underneath their protective scales, before dying. Scale insects can be divided into two types: the soft scale species within the Coccidae family and the armoured scale in the Diaspididae family. Some of the more common scale insects encountered are armoured scale species such as orchid scale (*Diaspis boisduvalii*) or cactus scale (*D. echinocacti*) and soft scale ones like hemispherical scale (*Saissetia coffeae*) or brown soft scale (*Coccus hesperidum*).

Scale insects can be very difficult to control. The most effective way to manage a scale infestation is by regularly using horticultural soaps and oils (see How to treat plants with insecticidal soaps, invigorators or soils, page 160), which will reduce the crawlers. Manual removal of adults can be achieved using a small paintbrush dipped in 70 per cent alcohol. A strong mist setting on an irrigation lance, used like a power washer, can dislodge them too.

Top, left: Large population of armoured cactus scale.
Top, right: Black soft scale adult.
Above: Hemispherical scale adult.

Like their close relative the mealybug (see page 140), soft scale species produce honeydew, which attracts ants; these help spread and protect scale insects from horticulturally beneficial insects. (Armoured scale species do not produce honeydew.) Reducing the ant population (see page 132) is essential in maintaining your biological control programme using nematodes such as *Steinernema feltiae,* ladybirds in the genus *Chilocorus* and lacewings in *Chrysoperla*.

Spotting new populations of scale insects early will go a long way to manage this pest, as will separating out infected plants to minimize their spread. Manually remove them when first spotted; then monitor the plants closely afterwards for new infections. Where there are large outbreaks it can be difficult to tell which scale insects are alive and which are dead but still persisting on the plant. Consider pruning out heavily infested stems and leaves, if possible, and then brush off as many of the dead scale as possible. This will make controlling the remaining population easier. Consider moving pot plants with scale away from unaffected plants, and if heavily infested think about replacing the plant or propagating a new one.

Red spider mites *Tetranychus urticae*

PLANTS AFFECTED

Wide range of indoor plants including cacti, orchids, Marantaceae, angel wings (*Caladium*) and *Alocasia*

DAMAGE CAUSED

Discoloration or speckling of the leaves and webbing over a leaf; weakened plants

Red spider mites, also known as two-spotted mites, are tiny arachnids, less than 1mm/½sin long, with two characteristic spots. They are cell-content-feeding insects thriving in dry warm environments. Under favourable conditions, small unnoticed populations can quickly become large infestations before they are spotted. Females, which live for four weeks as adults, can lay a hundred eggs or more. Progression from eggs to adults takes twenty days under optimum temperatures.

These mites can cause severe weakening of plants, leading to leaf drop, distortion and sometimes death. In cacti they can cause irreversible scarring in the apex of the plant. When conditions become unfavourable, such as cooler temperatures or reduced food supply, spider mites will move away from the plants to find a safe place, turn bright red and go into diapause (insect hibernation) while they wait for better conditions to start re-infecting the plant. However, in a greenhouse environment red spider mites can be active all year.

Top, left: Large infestations
of spider mite with
characteristic webbing.
Top, right: Adult red spider mite.
Above: Typical red spider
mite symptoms.

Keeping plants healthy, well fed and watered in favourable conditions will reduce instances of red spider mite. Plants grown too hot and dry will be a magnet, so regularly spraying over foliage to maintain humidity of more than 60 per cent will help discourage them. Good husbandry in the greenhouse is important: pick up and discard fallen leaf litter; and clean and disinfect staging, particularly in winter, to reduce overwintering mites.

There are several biological controls that can be very effective at maintaining low numbers of red spider mite. To use them effectively they should be released before these mites become a problem (see How to use biological controls, page 90). Predatory mites such as *Hypoaspis*, *Neoseiulus cucumeris* and *Phytoseiulus persimilis* can be very efficient if scattered through the foliage early in the growing season. Predatory midge *Feltiella acarisuga* is good at finding red spider mite populations that are difficult otherwise to reach or spot.

Horticultural oils and soaps can be very helpful if used frequently, and good coverage is achieved when applying (see How to treat your plants with insecticidal soaps, invigorators or oils, page 160). Such sprays will reduce the mobility of red spider mites, block the breathing pores and so reduce numbers.

Carnation tortrix moths   *Cacoecimorpha pronubana*

PLANTS AFFECTED
Wide range of temperate
plants in greenhouses

DAMAGE CAUSED
Damaged growing points
with leaves/petals 'sewn'
together

Above: Adult carnation
tortrix moth.

Carnation tortrix moths can be found within and outside the
greenhouse. Females lay up to 700 eggs on upper leaf surfaces
from spring and can produce multiple generations during the
year. When the bright green caterpillars hatch about ten days
later, they move towards the tip of the branch, where they sew
two leaves together before feeding. Once encased they are well
protected from predators, making control at this point tricky.
After pupating, small, light brown moths, with darker
banding, emerge.

Monitoring for this tortrix moth is invaluable, and well-
timed treatments shortly after hatching can be very effective.
Manually squash or prune any sewn leaf tips. Set up pheromone
lures in delta traps (see How to catch pests with sticky friends,
page 138) to attract and catch males. Ultraviolet light traps in
the greenhouse can also help catch tortrix moths but, as with
the pheromones, be careful not to
attract more moths through the vents.

Biological controls include the
nematode *Steinernema carpocapsae*
and *Trichogramma* parasitic wasp
species (see How to use biological
controls, page 90). Both can be very
effective; however, timing is crucial
– releasing *Trichogramma* after the
tortrix eggs have hatched or
applying nematodes once the
caterpillars are sewn into their
protective casing will not work.

Slugs and snails Gastropoda class, Mollusca phylum

PLANTS AFFECTED
Wide range of indoor
and outdoor plants
DAMAGE CAUSED
Rasping and holes in
the leaf surface

Slugs and snails are divided into two groups: snails have a defined shell into which they can retract; slugs have a very reduced or no shell at all. They both thrive in moist conditions, making greenhouses perfect environments for them, especially as they are protected from natural predators like birds, frogs and hedgehogs. They lay clusters of jelly-like eggs, each only 2–3mm/$\frac{1}{12}$–$\frac{1}{8}$in diameter, often in the soil or leaf litter.

Slugs and snails are often nocturnal feeders, staying well covered during the heat of the day. When they emerge from their hiding spots, they leave characteristic slime trails – often on the tastier vegetable plants in the garden or greenhouse.

If left, these molluscs can quickly become a problem, causing havoc in the greenhouse especially when there are young seedlings. However, it is worthwhile remembering that, although some slugs and snails feed on your plants, they also play an important role in the ecosystem by breaking down organic matter and helping return nutrients to the soil. Some slugs like the leopard slug (*Limax maximus*) can even act as a carnivore by devouring other slug and snail species.

For some ways to minimize problems from these molluscs see How to cope with slugs and snails, page 150.

Top: Adult snail.
Above: Adult slug.

How to cope with slugs and snails

Slugs and snails are most active at night, so difficult to spot during the daytime when they are often hiding under pots or in cracks and crevices. They are especially prevalent from spring to late autumn. They go into winter hibernation, when there is less food and temperatures are cold. During this time, they often cluster together under rocks and logs or in empty flower pots in the greenhouse. Seeking out these hibernation spots is a great way of reducing the population before spring.

Try to fill in gaps in the greenhouse, thereby preventing these molluscs from entering. Keep your greenhouse clean and clutter-free, reducing areas in which they can hide. Check regularly under pots, or visit after dark and manually remove any you spot. When starting vegetables in the greenhouse ensure you have well-established plants before planting them out in the garden (see How to grow strong plants from seed, page 40). Once plants are larger and more established, slug damage is less detrimental to them. Sowing a few more plants than needed will allow for losses to slugs and snails, too.

Encourage wildlife into the garden to reduce the slug population outside the greenhouse, which will mean fewer trying to get in. Providing habitats for birds, frogs, hedgehogs and beetles will help maintain an ecological balance in the garden, thereby preventing slugs and snails from becoming a problem.

In the garden and greenhouse, nematodes such as *Phasmarhabditis hermaphrodita* can be effective at controlling a range of slugs and snails. They are tiny, worm-like creatures, less than 0.5mm/1/50in long, naturally occurring in

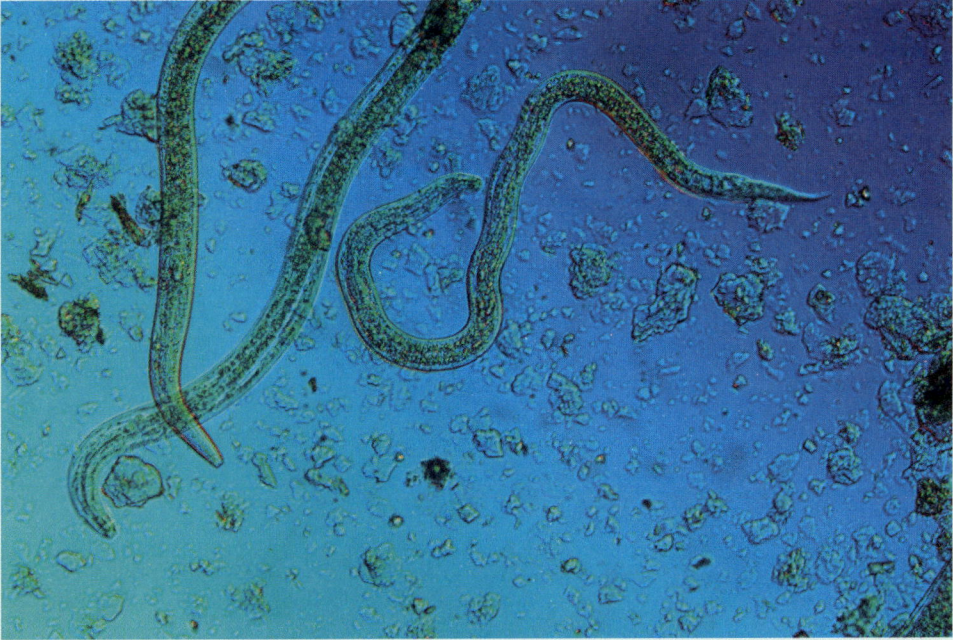

soil. They will seek out slugs and snails and enter the slug through its breathing hole. Once inside, they feed on the slug and infect it with a bacterium that will ultimately kill it. They are best introduced in spring, when the temperature is above 5°C/41°F, or in autumn.

To use these nematodes effectively they should be applied in early morning or in the evening, when they are less likely to dry out and be exposed to high light levels. Add the nematode sachet to a jug of lukewarm water and stir well; then let them sit for 5 or 10 minutes. Add the nematode liquid to a watering can with the remaining water required (see manufacturer's instructions). Apply to the area where slugs and snails are present, agitating the liquid frequently, as these nematodes will settle at the bottom of the watering can very quickly. If you use a sprayer, be sure to remove any filters (see How to use biological controls, page 90).

Left: Snails often overwinter around terracotta crocks.
Above: *Phasmarhabditis hermaphrodita* nematodes.

Thrips Thysanoptera order

PLANTS AFFECTED
Wide range of plants
including orchids, succulents
and other houseplants

DAMAGE CAUSED
Silvering distorted leaves,
flowers and fruit

Top: Typical thrips damage.
Above: Adult glasshouse thrips.

There are over 7,000 species of thrips described to science;
fortunately, not all of them are sap feeders. The few species that
feed on plants can not only cause significant physical damage,
but they are also a major vector of plant viruses.

Thrips are often golden brown to black. Adults are thin,
winged insects usually no more than 2mm/¹⁄₁₂in long. Juveniles
are similar in appearance, but wingless.

To discourage them, open greenhouse vents on the opposite
side to the prevailing wind so thrips are blown over the greenhouse.
Remove flowers or prune plants if heavily infested.

Sticky traps are an effective way to monitor and reduce
populations, as are pheromone lures for some species (see How
to catch pests with sticky friends, page 138). Suitable beneficial
insects include *Hypoaspis, Atheta coriaria* and beneficial
nematodes for species that pupate in the soil (see How to use
biological controls, page 90). Predatory mites such as *Neoseiulus
cucumeris, Amblyseius montdorensis, A. swirskii* and *Orius* species
can also be efficient. Alternatively, try horticultural soaps and oils
(see page 160).

Fusarium wilt *Fusarium oxysporum*

PLANTS AFFECTED
Wide range of ornamental plants

DAMAGE CAUSED
Stunted growth; yellowing and wilting leaves; red staining of xylem, leading to stem and root rot

Top: Typical fusarium wilt damage.
Above: Fusarium wilt symptoms.

This soil-borne fungus is a weak pathogen and is sometimes saprophytic. As a result, fusarium wilt infects only stressed plants through damaged or weak roots. The fungus infects the xylem within the base of the plant, from where it moves upwards through the plant. The plant responds by trying to block off the pathways of infection, preventing further spread through the xylem. This causes reduced ability to transport water to the leaves, resulting in the typical wilt response.

Fusarium wilt can be more prevalent during warmer summer months, when plants can experience heat and drought stress. Keeping your plants healthy will reduce cases of it, as will growing disease-resistant cultivars. Sadly, there is little that can be done to treat infected plants. If you suspect fusarium wilt, destroy infected plants; sterilize tools, pots and greenhouse staging (see How to clean tools and other equipment, page 104); and avoid reusing any potting compost.

Be warned: fungal pathogens can lay dormant for many years in the form of resting bodies known as chlamydospores, so fusarium wilt may re-emerge.

Grey mould *Botrytis cinerea*

PLANTS AFFECTED
Wide range of ornamentals
and greenhouse crops

DAMAGE CAUSED
Furry grey mould on the
stem, leaves, fruit or flowers

Grey mould is one of the most common plant pathogenic
fungi found in greenhouse environments, particularly where
there is humidity higher than 90 per cent on the leaf surface
and a lack of airflow. Grey mould is difficult to control
because of its various modes of attack, diverse hosts and
survival in multiple forms. It prefers temperatures of 18–24°C
(65–75°F) and so becomes more problematic in autumn
and winter. In ideal conditions, grey mould can develop
quickly on dead organic material, initially feeding on dead
plant material but later becoming a plant pathogen entering
healthy tissue through dying leaves or wounds. It will infect
plants that are already weakened rather than healthy
specimens and can lead to the plant dying.

Preventing outbreaks is important
as there is little in the way of control
once grey mould has infected a plant.
Feed plants well so they have better
resistance to infection. Good hygiene
is also important in managing grey
mould. Remove dead leaves and
flowers regularly. Ensure there
is airflow around the plants,
spacing them out and pruning
where needed. Ensure
your plants are not over-
or underwatered.

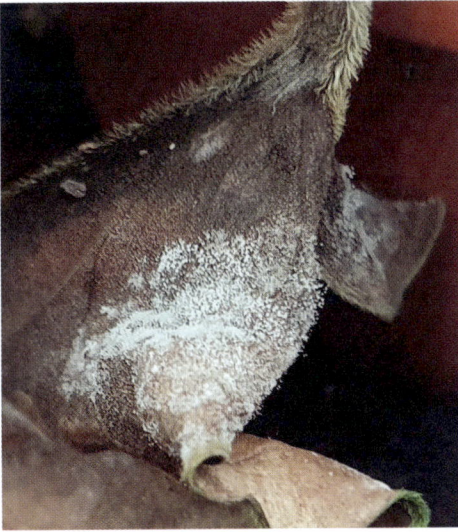
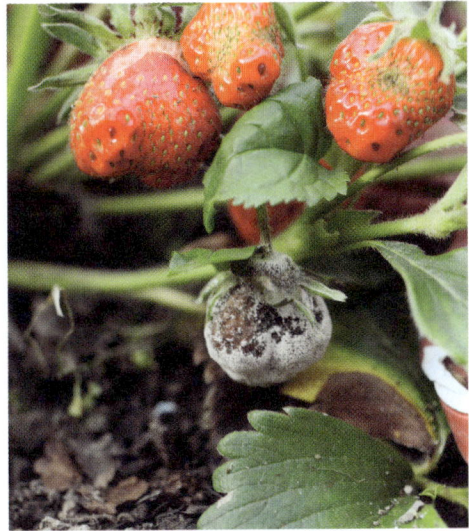

Above, left: Grey mould
symptoms.
Above, right: Damage from
grey mould on strawberries.

When watering, try to avoid getting the leaves wet, particularly
on dull days or in the evening, when they will not dry out quickly.
In unfavourable conditions, this fungus will be less active;
however, it can become dormant, forming sclerotia (resting
bodies) in the dead tissue, which will become active again when
favourable conditions return.

Grey mould can be particularly problematic in propagation
environments, causing cuttings to fail and young seedlings to
die. Clean propagation areas regularly, removing failed cuttings
and abscised (broken, cut off or deciduous) leaves as soon as
possible. Spray propagators with a horticultural disinfectant
before use (see How to clean tools and other equipment, page
104), and always choose clean sharp secateurs and knives when
taking cuttings.

Beneficial fungi and bacteria such mycorrhizal inoculates
and biological stimulants can act antagonistically, inhabiting
the space where harmful fungi would grow and producing
anti-fungal compounds, thereby reducing competition by
pathogens. Biological fungicides based on denatured yeast are
becoming available. When sprayed on the plant, these products
mimic a fungal attack, triggering the plant's natural defences
before the pathogen is present. All these products are best used
as a preventative rather than a curative.

Sooty mould Various, including *Cladosporium* and *Alternaria* species

PLANTS AFFECTED
Wide range of plants attracting honeydew pests or producing sticky plant residues

DAMAGE CAUSED
Leaves and stems; weakened plant

Sooty mould is the collective term for several ascomycete fungi (including the genera *Cladosporium* and *Alternaria*) that produce black mats on the living leaves and stems of plants. It is often associated as a secondary symptom of honeydew-producing insects like scale insects (see page 144), mealybugs (see page 140), aphids (see page 134) and whitefly (see pages 38 and 166). Sooty mould grows on sugar residues left on plants: for example, honeydew or, in some cases, nectar or plant exudates.

Above, left: Sooty mould symptoms.
Above, right: Black mats of sooty mould on *Camellia*.

Unlike pathogenic fungi, species that produce sooty mould do not infect the plants or cause immediate harm to them. However, as well as being unsightly, widespread sooty mould can reduce the plant's ability to photosynthesize. When combined with a plant weakened by pest infestation, the impeded ability to photosynthesize can be detrimental.

Controlling these moulds is simple: remove the sugar source. This may entail wiping leaves with warm water or spraying them to wash away sugar residues. Where the residue is due to honeydew, treating the pest infestation is crucial. Insecticidal soaps such as a plant invigorator are a good option because the soap will not only help wash away sugar but will also reduce the pest causing the problem (see How to treat your plants with insecticidal soaps, invigorators or oils, page 160). Sooty mould is often more prevalent where humidity is above 60 per cent and there is a lack of airflow. Improving air movement and reducing humidity will help to minimize incidents.

Bacterial leaf spots   *Pseudomonas* and *Xanthomonas* species

PLANTS AFFECTED
Calathea, Begonia, busy
Lizzies (*Impatiens*), *Monstera*
and *Pelargonium*
DAMAGE CAUSED
Spotted leaves; chlorosis;
leaf drop

Top: Typical bacterial leaf
spot symptoms.
Above: Bacterial leaf
spot damage.

Bacterial leaf spots can be very similar to fungal infections although less common. Bacterial spotting is usually between the veins and often starts at the leaf margins. The spots often have a wet appearance with bacterial ooze, and can have a bright yellow halo. Spotting is often more angular than fungal leaf spots (see opposite).

Bacteria usually enter plants through cuts, wounds and stomata (pores). The disease can spread rapidly under cool wet conditions. Ensure irrigation water is free of debris and avoid wetting leaves, especially if they will remain damp into the evening. Reduce humidity and allow good airflow around plants.

Managing bacterial disease starts with good hygiene: keep tools and growing spaces clean (see How to clean tools and other equipment, page 104); remove fallen debris; and cut and dispose of infected leaves from the plants.

Because the bacteria are inside the tissue of the infected plant there is very little that can be done to cure an infected spot; therefore, preventing re-infection is the best solution. Bactericides can be beneficial; however, they will control only surface bacteria. Beneficial bacteria such as *Bacillus amyloliquefaciens* colonize the plant surface inhibiting pathogens. After a disease incident, sterilize surfaces, pots, hosepipes, watering cans and tools.

Fungal leaf spots *Alternaria, Septoria* and *Colletotrichum* species

PLANTS AFFECTED
Wide range of greenhouse
and houseplants
DAMAGE CAUSED
Brown spots

Many fungal species cause leaf spots. Identifying which species is infecting your plant can be difficult without sending infected material off for analysis. Fungal leaf spots can be confused with bacterial leaf spots (see opposite), nutrient deficiencies or environmental conditions. Fungal spots are circular or oblong and can be black, purple, brown or yellow in colour. Spots can extend across the veins; they may have concentric rings or visible hyphae.

To deal with fungal leaf spots, it is important to understand the disease triangle (see page 12). If you have plants that are very susceptible to fungal diseases, consider growing less susceptible species or varieties. Avoid wetting the foliage especially if it won't dry quickly before sunset. Use fans, space plants correctly and prune, to allow air movement through the plant.

Typical fungal leaf spot symptoms.

Virus

PLANTS AFFECTED
Wide range of indoor
and outdoor plants
DAMAGE CAUSED
Mosaic, streaks or ring
patterns on the leaves;
stunted distorted growth
and reduced vigour

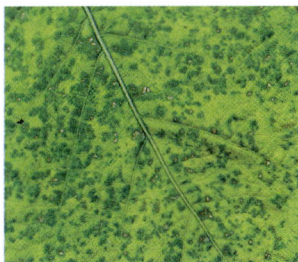

There are many viruses that can infect plants. Some cause significant damage while others may even be encouraged, as is the case for some variegated plant cultivars. Viruses work by 'hijacking' the cell and are then replicated through cellular division. There is very little that can be done to cure a plant of a virus.

Viruses are transmitted during vegetative propagation or via sap-sucking insects. To prevent them, disinfect your secateurs regularly (see How to clean tools and other equipment, page 104), control sap-sucking insects and remove plants including weeds carrying viruses.

Typical symptoms of a virus attack.

How to treat your plants with insecticidal soaps, invigorators or oils

If you do need to spray a pest-infested plant, use environmentally friendly products such as horticultural soaps, invigorators and oils. These have a physical mode of action, breaking down an insect's waxy layers and blocking its spiracles. They also leave no chemical residues, and the insect is less likely to develop resistance. Plant invigorators act as soaps but have the added advantage of acting as a foliar feed, thereby stimulating plant growth.

When treating a pest, first identify it and then select a suitable product. Remember not all treatments are effective for every pest. It's important to read the label of any product you are using. This will tell you what the active ingredients are, what the application rates are if using a concentrate, and what health and safety precautions to take. To ensure you have enough spray to cover all the foliage to the point of run-off, calculate the amount of spray you need.

Before mixing up your spray, check that you have put on all the required personal protection equipment, such as nitrile gloves and eye protection. It is good practice to mix your spray in a plastic tray; this way if there are any spillages, they will be contained. Then fill your sprayer with half the water volume

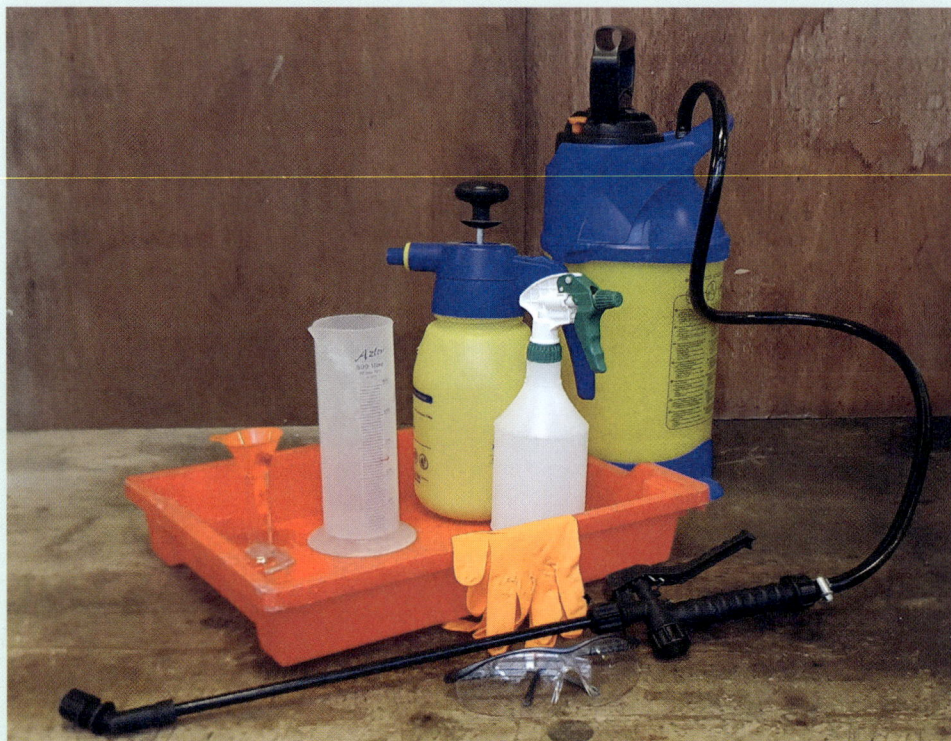

required. With the lid sealed, gently upturn the concentrate bottle a few times before measuring out the required product and adding it to the sprayer.

Use some of the remaining water to rinse the measuring cylinder into your sprayer, until it is thoroughly clean. If you finish the concentrate, triple-rinse the empty bottle into the spray, too. Add the remaining water to the sprayer and agitate the liquid to ensure it is well blended.

Spray your plants in the afternoon. This will help to avoid spray damage from sunlight, and it will allow the leaf surfaces to stay wetter for longer, increasing the spray efficacy. In a greenhouse, damp down the floor to enhance the humidity, helping to slow the drying time of the spray. Close the shading if available, to reduce the risk of scorch, turn off any fans and close the vents to exclude any bees or other pollinators.

Having first tested the mixture on a leaf or two, spray both the tops and undersides of the leaves, to get good coverage. Wet the surfaces until the point of run-off. Spray all the plants in the vicinity, paying particular attention to any you know have pests. Use up any excess spray, and triple-rinse out your sprayer after application.

After spraying, monitor the plants. Repeat applications a week later, so any newly hatched individuals are treated. Pests with longer life cycles may require multiple applications.

Opposite: Among essentials needed for spraying pests is a tray to contain any spills when filling the sprayer.
Above: When spraying, ensure the entire leaf including the underside is well covered, here with horticultural soap on a *Caladium* leaf.

Powdery mildew *Erysiphe, Podosphaera, Oïdium* and *Leveillula* species

Powdery mildew is caused by several different fungal species in the order Ascomycota. Its fungal spores germinate rapidly in humidity above 70 per cent. Infection is often found first on the underside of the leaf, then the white powdery coating spreads to the rest of the plant. The optimal condition for powdery mildew is a temperature of 18–25°C/65–77°F and high humidity for spore germination, followed by dry conditions.

To reduce incidents of powdery mildew, ensure good air movement and adequate light, and avoid fluctuations in humidity and wetting the foliage, especially if the plant is likely to stay wet overnight. Water plants well during dry sunny periods – wetting foliage in these favourable conditions can reduce mildew. Keep your growing space and tools clean (see How to clean tools and other equipment, page 104), and remove fallen and infected leaves.

To make your plants more resistant to attack, feed them with silicon to strengthen cell walls.

Top: Powdery mildew
symptoms.
Above: Rose buds covered in
powdery mildew.

Downy mildew *Peronospora, Plasmopara* and *Bremia* species

PLANTS AFFECTED
Wide range of ornamentals
and greenhouse plants

DAMAGE CAUSED
Whitish grey patches on the
underside of the leaf causing
blotching on the upper side

Top: Leaf damage from
downy mildew.
Above: Downy mildew
symptoms.

This is a similar pathogen to powdery mildew (see opposite); in that it infects the underside of the leaf, however, it causes discoloration on its upper side. It is an oomycete, not a true fungus, and favours temperatures of 15–21°C/59–70°F with prolonged periods of humidity, particularly in spring and autumn. This infection is less likely to be problematic outdoors during dry summer conditions; however, it can persist year-round in greenhouses with humidity above 85 per cent.

Downy mildew releases its spores at night and requires moisture to germinate, so to prevent it avoid overcrowding; encourage air movement; reduce humidity if possible; and do not leave foliage wet overnight. Good hygiene such as cleaning benches, tools and pots is important (see How to clean tools and other equipment, page 104), as is the removal of fallen leaves and infected material.

Downy mildew can produce oospores, which can lie dormant in the soil and act as the initial infection. Therefore, avoid composting infected plants and use good-quality potting soil. Because downy mildews can be host-specific, consider growing different species if the outbreak is severe.

Root rot

Fungi: *Fusarium, Rhizoctonia* and *Phoma* species
Oomycetes: *Pythium* and *Phytophthora* species;
Bacteria: *Pectobacterium* and *Dickeya* species

PLANTS AFFECTED
Wide range of houseplants
including cacti and succulents

DAMAGE CAUSED
Yellowing or browning leaves;
wilting; leaf drop

Root rot can be caused by several pathogens, and it is often linked to overwatering. Constantly damp potting mixes become anaerobic, reducing plant health and making plants more vulnerable to root-rot pathogens. One of the challenges of root rots is that the first sign is often wilting. This occurs if there is a lack of water in the pot, as well as if the plant has no roots to take up the water. Unfortunately, the gardener's automatic response to a wilting plant is to increase watering, but, if the wilt is caused by rot, additional watering will make the situation worse.

Preventing root rot starts with cleanliness in the growing space and of tools, as they are often spread through water sources and potting media (see How to clean tools and other equipment, page 104). Grow your plants in a good-quality potting media that is pathogen-free and suitable for the species or variety. Your water source must be of a good standard, too. Adequate drainage and appropriate growing conditions – correct light, temperature, humidity and nutrients – will keep your plants happy.

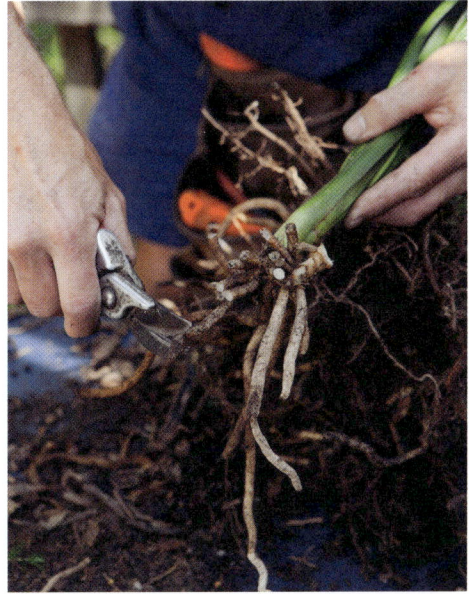

Above, left: Wilting and damaged leaves might indicate root rot.
Above, right: Dealing with root rot.

If you suspect a plant has root rot, remove it from the pot, then discard the soil and inspect the roots. Plants with rot have a strong odour, discoloured root tissue and may have white threads (if it is fungal) or slimy ooze (if bacteria have caused the infection). Remove signs of rot by cutting back to healthy root tissue; always sterilize secateurs between cuts. Treat the cut ends with powdered charcoal or sulphur dust, then allow the plant to dry out before repotting into fresh compost. Depending on the plant it may be useful to shorten the top growth, to reduce transpiration. After repotting, water cautiously or leave dry for a week or so if the plant is a succulent. Consider taking cuttings to start a new plant.

Encourage antagonistic fungi and bacteria in the growing media to make it harder for pathogens to compete. Several soil microbial products, such as mycorrhizal inoculants, can help create a balanced soil biology.

Glasshouse whitefly *Trialeurodes vaporariorum*

PLANTS AFFECTED
Temperate greenhouse and houseplants, especially those with slightly hairy leaves

DAMAGE CAUSED
Stunted growth; yellowing moulted leaves with sticky residue, leading to sooty mould

These vectors for viruses produce honeydew, which can attract ants (see page 132) and lead to sooty mould (see page 156). They are most often found on the undersides of leaves, appearing only when disturbed. The adults, 2mm/1⁄12in long, are winged and white to pale yellow in colour. Their eggs are laid in a characteristic horseshoe pattern. These hatch after a few days as crawlers and move around the host plant to find a suitable feeding spot. They then go through two moults, where they lose their legs and become immobile. At the fourth larval stage the whitefly turns into a filamentous, scale-like pupa before emerging as a winged adult ten days later. Progression from egg to egg-laying adults takes 35–40 days.

These pests can easily go unnoticed until they reach large populations. Spotting them early is important to effectively manage them. To do this, hang yellow sticky traps (see How to catch pests with sticky friends, page 138) around susceptible plants. Also, adopt good cultural practices in and around your greenhouse: keep the area clean and weed-free; and ensure good airflow around plants.

Above, left: Adult whiteflies.
Above, right: Different developmental stages of whitefly.

Populations can be reduced by taking your plant outdoors and spraying it with the fan or mist setting of on your watering lance, thereby dislodging the adults. Horticultural soaps and oils (see How to treat your plants with insecticidal soaps, invigorators or oils, page 160) can also be effective if repeated regularly.

Several beneficial insects can be used to control whitefly. *Encarsia formosa* is a parasitic wasp, which lays a single egg inside the juvenile whitefly, turning it black (see page 21). This wasp seeks out the whitefly by smell, which makes it less effective when searching for whitefly among fragrant plants. *Macrolophus* is a small green bug that eats whitefly. It lays its eggs inside the leaf, causing minimal damage, and prefers slightly hairy leaves, like sage (*Salvia*). So, to establish these beneficial bugs ensure you have a suitably fragrant-leaved plant among your plants. Other biologicals for whitefly include *Eretmocerus eremicus* and *Amblyseius swirskii*.

Glossary

ABSCISED When a plant organ such as a leaf naturally comes away from the main plant.

ABSCISIC ACID A plant hormone involved in plant development such as seed germination and abscission, and in stress responses.

ALLELOPATHY The negative or positive effect on the growth and development of surrounding organisms when a plant releases toxic chemicals.

AMBROSIA FUNGI A group of fungi that has a symbiotic relationship with some beetles, known as ambrosia beetles.

ANAEROBIC Without oxygen.

ASCOMYCETE FUNGI (aka sac fungi) A group of fungi that have a tiny, sac-like sexual structure where their spores are formed.

BACTERIA Tiny, single-celled organisms.

CALYX Collective name for the sepals, which make up the outer whorl of a flower.

CAMBIUM A tissue layer of actively dividing unspecialized cells in a plant stem that can specialize once they divide.

CANKER An area of dead, often sunken stem issue.

CATCH CROP A fast-growing plant that can be sown and harvested between main-crop rotations; it helps to reduce nutrients leaching from the soil.

CHLAMYDOSPORE The thickened-walled resting spore of some fungus.

CHLOROSIS A lack of chlorophyll in green plants causing leaves to turn yellow because of many factors, including nutrient deficiency, pests and diseases and also the normal life cycle.

CONIDIOSPORES Non-mobile, asexual spores produced by some fungi.

DETRITIVORE An organism that feeds on dead organic material.

DIAPAUSE A period of dormancy in an organism's life cycle.

EELWORM (q.v.) nematode.

ENDOPARASITIC WASP A parasitic wasp that lays its eggs in a host. When the eggs hatch, the larvae live on the host and ultimately kill it.

ENDOPHYTE Usually a fungus or bacterium that lives in the tissues of a host plant.

ENTOMOPATHOGENIC FUNGI They live in soil and infect insects, eventually killing them.

EPIPHYTE A plant that grows on another plant rather than in the soil and gets water and nutrients from the air.

FRASS Insect excrement.

FUNGUS An organism that produces spores and absorbs its nutrients from decaying or living plants.

GALL A swelling on external plant tissue, often caused by insects.

GIRDLE (aka ring-barking) To remove the outer layer of tree bark either purposely or through damage.

GUMMOSIS The oozing of sap from a tree as a result of wounding.

HAULM Potato stem.

HYPERSENSITIVITY PNEUMONITIS Inflammation in human lungs caused by an inhaled allergen.

HYPHAE Branched tubes of fungi, which secrete enzymes and absorb nutrients; they are collectively known as mycelium.

INSTARS Stages in an insect's life cycle.

LEAF MINE A tunnel inside a leaf left behind by larval stages of some insects.

LENTICEL A pore in the skin of a plant, which allows gas exchange between the underlying tissue and the atmosphere.

LESION Damaged tissue caused by various factors including wounds.

MESOPHYLL TISSUE (in a leaf) The cells in a leaf where photosynthesis (q.v.) occurs.

MYCELIAL CORD (q.v.) rhizomorph.

MYCELIUM (q.v.) hyphae.

MYCORRHIZAL FUNGI The beneficial fungi that associate with plant roots by expanding the plant's network.

NECROTIC Dead or dying tissue.

NEMATODE (aka roundworm and eelworm) Incredibly small worms that often live in the soil and can be used as a pest control, but some are pests, such as potato eelworm.

OOMYCETE (aka water-mould) An organism with spores and filaments (similar to a fungus, q.v.) often causing plant diseases such as blight (see page 24).

OOSPORES The sexual spores

of some fungi (q.v.) and oomycetes (q.v.) with thick walls allowing them to survive for long periods of hibernation.

PARTHENOGENESIS Asexual reproduction.

PATHOGEN An organism that causes disease.

PEA STICK A twiggy-topped hazel (*Corylus*) stem harvested to use as a support for a pea plant.

PETIOLE The part of a plant that attaches the stem to the leaf.

PHLOEM Vascular tissue that moves soluble organic compounds around a plant; these are primarily sugars and the foods created during photosynthesis (q.v.).

PHOTOSYNTHESIS A reaction in plant cells using water, carbon dioxide and energy from the sun to make glucose and oxygen, which is used as a food source by the plant.

POME FRUIT Fleshy fruit (such as an apple or pear) derived from the tree's female reproductive organs and having a core that contains seeds.

PROBOSCIS (of an insect) The long mouthpart used to suck up food.

PRONOTUM The plate-like structure that covers an insect's thorax (q.v.) directly behind its head.

PROTIST A single-celled organism belonging to the Protista kingdom, which includes most algae, protozoans and slime moulds.

PUPATE To develop into a pupa, the stage in the life cycle of some insects, between immature larval and adult stages.

PUSTULE An eruption on the surface of a plant often caused by rusts.

RHIZOMORPH (aka mycelial cords) Plant-root-like structures that are the hyphae (q.v.) of some fungi.

RING-BARKING (q.v.) girdle.

ROOT EXUDATE A fluid released from plant roots to stop harmful organisms.

ROUNDWORM (q.v.) nematode.

SAC FUNGI (q.v.) ascomycete fungi.

SAPROPHYTIC Obtaining nourishment by feeding on dead or dying organic material.

SAPROTROPH An organism that feeds on dead or dying material.

SCLEROTIUM A compact mass of hardened mycelium ((q.v.) hyphae) that contains food stores and helps fungi survive when conditions are not ideal.

SPIRACLE (of an insect) A respiratory opening on the thorax (q.v.).

SPORE A reproductive unit in some fungi.

STEM BLEEDS Vertical weeping lesions (q.v.) on an oak (*Quercus*) tree that indicate the tree is affected with acute oak decline AOD (see page 112).

STOMATA Pores in the surface of a leaf or stem, which allow gas movement in and out of the plant.

STROMA (in fungi) A dark hard mass of tissue formed of hyphae (q.v.) that contain spore-bearing structures.

STYLET Part of the mouthpiece of some insects, which acts like a hypodermic needle to allow the sucking of food such as sap.

THORAX (of an insect) The middle of three sections of an insect body, where the legs and wings are attached.

VECTOR An organism such as an insect that transfers a disease from one host to another.

VOLUNTEER POTATOES Those that were not harvested in the previous season and sprout again the following year.

WATER-MOULD (q.v.) oomycete.

WITCHES' BROOM (in a tree) An area of abnormal dense twig and branch growth from a single point, usually caused by pests and diseases.

XYLEM Vascular tissue in plants that transports water and nutrients from the roots upwards.

Index

Page numbers in **bold** indicate 'How to' sections; page numbers in *italics* indicate illustrations (not including main articles on specific pests which all have illustrations)

Index

glasshouse leafhoppers 137
Gloxinia 136
greenfly 134–5
grey mould 154–5

H

Hauptidia maroccana see glasshouse leafhoppers
hawthorn 78–9
hedgehogs 33
heliothrip *152*
hemlocks 106–7, 122–3
Heterorhabditis bacteriophora 63, 91
Hibiscus 136
holly 98
honey fungus 100–1
honeydew 132, 134, 141, 145, 156, 157
horse chestnuts 98, 115
houseplants 130–68
hoverflies 53, 134
hyacinths 86–7
Hydrangea 86–7
Hymenoscyphus fraxineus see ash dieback
Hypoaspis 143, 147, 152

IJK

Itersonilia perplexans see canker, parsnip
kale 38

L

lacewings *15*, 19, 134, 137, 141
ladybirds 19, *21*, 134, 141, 145

larches 97, 106–7
Lasius niger see ants
lavenders 98
lawns 63, 85
leaf miners: chrysanthemum 70–1; horse chestnut 115
leaf spots: bacterial 158; fungal 159; rhubarb 59
leafhoppers 93, 137
leeks 48, 49
leopard slug 149
lettuces: growing from seed *40*; sacrificial 52
Leveillula spp. *see* mildew, powdery
Lilioceris lilii see beetles, lily
Limax maximus see slugs, leopard
Lymantria dispar see moths, European gypsy

M

Macrolophus 167
maples 98, 102
Marantaceae 146–7
marigolds as companion planting *53*
mealybugs *8*, 132, 140–1
methylated spirits 141
Microdochium nivale see fusarium patch
midge, predatory 147
mildew: downy 163; powdery *13*, 84, 162
millipedes 142
mint as companion planting 27, 47, 52
mites 68–9; broad 135;

predatory 136, 143, 147, 152; red spider 146–7; two-spotted 146–7
moles *18*
molluscs *see* slugs; snails
Monilinia spp. *see* brown rot, fruit
Monomorium minimum see ants
Monstera 158
mosaic virus, daffodil 85
moths: box tree 66–7, 91; carnation tortrix 148; codling 28–9; European gypsy 96; horse chestnut leaf miner 115; leek 49; oak processionary 110–11; pea 39; tortrix 148
mulching 116, 127
mycorrhizal inoculants 165
Mycosarcoma maydis see corn smut
Myzus cerasi see blackfly, cherry

N

Narcissus yellow stripe virus 85
nasturtiums 37, 52
nematodes 91; carrying viruses 51; eelworms 86–7; friendly 20, 38, 63, 67, 73, 143, 145, 148, 150–1, *151*, 152
Neoseiulus cucumeris 147, 152
netting 17, 27, 33, 37, 47, 65, 71, *75*
nettles 19

O

oaks 96, 97, 110–11, 112–13, 114
Oïdium spp. *see* mildew, powdery
onions 48, 49
oomycetes *see* root rot
Ophiostoma novo-ulmi see Dutch elm disease
orchids 144–5, 146–7, 152
oriental chestnut gall wasps 118
ornamental plants 62–93
Otiorhynchus sulcatus see vine weevils

P

parsnips 45, 46–7
passport labels 16
pears *12*, 28–9. 49, 58, 78–9
peas 39
Pectobacterium spp. 76–7; *see also* blackleg, potato; root rot
Pelargonium 136, 158
Peperomia 136
peppers 56–7, 59
Peronospora spp. *see* mildew, downy
Phasmarhabditis hermaphrodita 150, *151*
pheromone traps 29, 96, 138–9, 148, 152
Phlox 86–7
Phoma spp. *see* root rot
Phyllotreta see beetles, flea
Phymatocera aterrima see Solomon's seal sawfly

Phytomyza syngenesiae see leaf miners, chrysanthemum
Phytophthora see blight; root rot
phytoplasmas 93
Phytoseiulus persimilis 147
Pieris spp. *see* butterfly, cabbage white
pigeons 64–5
pines 97, 106–7, 108–9
planes 128
Planococcus spp. *see* mealybugs
Plasmodiophora brassicae see clubroot
Plasmopara spp. *see* mildew, downy
Platypus cylindrus see borers, oak pinhole
plums 49, 54–5, 124–5
Podosphaera spp. *see* mildew, powdery
Polyphagotarsonemus latus see mites, broad
poplars 96
potatoes 24–5, 43, 44
pruning for canker 125
Prunus 124–5
Pseudomonas syringae see canker, bacterial
Psila rosae see fly, carrot
Psylliodes see beetles, flea
Pyrethrum-based sprays for cypress aphids 103
Pyracantha see firethorn
Pythium 76–7; *see also* root rot

QR

quince 78–9
rabbits 72; protection from **74–5**
radishes 26–7, 47
Ramularia rhei see leaf spot, rhubarb
raspberries 98
red band needle blight 106–7
red spider mites 146–7
replant disease 84
Rhizoctonia spp. *see* root rot
Rhizoecus spp. *see* mealybugs
Rhododendron 80–1, 98
rhubarb 59
rocket 26–7
root collars 33, **34–5**
root fly: cabbage 32–3; carrot 46–7
root rot 164–5; phytophthora 98–9
roses 78–9, 82–3, 93, *162*
rove beetles 143
rusts *12–13*; rhubarb *see* leaf spot, rhubarb

S

Saissetia coffeae see scale insects
sawfly 88–9
scab: apple and pear 58; common 43
scale insects 132, 144–5
Sciurus carolinensis see squirrels, grey
sclerotinia 92
seed, growing from **40–1**
Seiridium cardinale see

Index

Author biographies

Hélèna Dove is the Kitchen Gardener at the Royal Botanic Gardens, Kew, where she focuses on experimenting with new crops and techniques, especially in the area of climate change. She trained in horticulture on the Historic and Botanic Gardening Training Programme, and this led to a passion for heritage vegetables. She also specializes in trained fruit and orchard management. She has written several books on vegetable gardening and the botany of the crops we use, including *The Kew Gardener's Guide to Growing Vegetables.*

Kevin Martin is Head of Tree Collections at the Royal Botanic Gardens, Kew. Charged with safeguarding the future of the collection amid the dynamic landscape of climate change, Kevin's passion lies in comprehending its impact on trees and pioneering methods to pinpoint species resilient to forthcoming environmental shifts. Currently pursuing a research master's degree at Lancaster University, he focuses on species distribution modelling and future climate projections. He aims to revolutionize tree selection strategies, ensuring that Kew remains at the forefront of botanical conservation.

Paul Rees is Tropical Nursery Manager at the Royal Botanic Gardens, Kew. He studied horticulture in South Africa and then worked for an indigenous nursery and landscaping company before starting his own landscape business, growing native plants and creating ecological gardens. He furthered his training at Kew, where he now grows and curates the succulent collections, manages the Tropical Nursery glasshouse and oversees Kew's Glasshouse Biological Control and Pest Management Programme. He has written *The Kew Gardener's Guide to Growing Cacti and Succulents.*

Until recently, **Polly Stevens** was Decorative Unit Manager at the Royal Botanic Gardens, Kew, where she supervised the Decorative Display team and redesigned the Parterre, seasonal bedding and borders. She completed a horticultural diploma with the Royal Horticultural Society, based at Wisley Gardens. This led to a year of study in the USA as the Royal Horticultural Society/Garden Club of America Interchange Fellow. She is now developing a private garden in Surrey, UK and is the author of *Kew Answers for Everyday Gardeners.*

Quarto

First published in 2026 by Frances Lincoln,
an imprint of The Quarto Group.
One Triptych Place, London, SE1 9SH,
United Kingdom
T (0)20 7700 9000
www.Quarto.com

EEA Representation, WTS Tax d.o.o., Žanova ulica
3, 4000 Kranj, Slovenia. www.wts-tax.si

A catalogue record for this book is available from the
British Library.

ISBN 978-0-71128-890-4
Ebook ISBN 978-0-71128-891-1

10 9 8 7 6 5 4 3 2 1

Design by Arianna Osti
Publisher: Philip Cooper
Editorial Director: Alice Graham
Senior Editor: Michael Brunstrom
Project Editor: Joanna Chisholm
Assistant Editor: Isabella Toner
Senior Designer: Isabel Eeles
Senior Production Controller: Rohana Yusof

Printed in Guangdong, China TT/Sep/2025

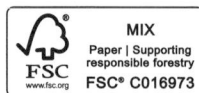

MIX
Paper | Supporting
responsible forestry
FSC® C016973